Buch Nr. 91/0651

Dieses Buch ist zurückzugeben bis zum

D1678118

REGULATIONS
FOR THE SAFE TRANSPORT
OF RADIOACTIVE MATERIAL

1985 Edition
(As Amended 1990)

The following States are Members of the International Atomic Energy Agency:

AFGHANISTAN	HOLY SEE	PERU
ALBANIA	HUNGARY	PHILIPPINES
ALGERIA	ICELAND	POLAND
ARGENTINA	INDIA	PORTUGAL
AUSTRALIA	INDONESIA	QATAR
AUSTRIA	IRAN, ISLAMIC REPUBLIC OF	ROMANIA
BANGLADESH	IRAQ	SAUDI ARABIA
BELGIUM	IRELAND	SENEGAL
BOLIVIA	ISRAEL	SIERRA LEONE
BRAZIL	ITALY	SINGAPORE
BULGARIA	JAMAICA	SOUTH AFRICA
BYELORUSSIAN SOVIET SOCIALIST REPUBLIC	JAPAN	SPAIN
	JORDAN	SRI LANKA
CAMEROON	KENYA	SUDAN
CANADA	KOREA, REPUBLIC OF	SWEDEN
CHILE	KUWAIT	SWITZERLAND
CHINA	LEBANON	SYRIAN ARAB REPUBLIC
COLOMBIA	LIBERIA	THAILAND
COSTA RICA	LIBYAN ARAB JAMAHIRIYA	TUNISIA
COTE D'IVOIRE	LIECHTENSTEIN	TURKEY
CUBA	LUXEMBOURG	UGANDA
CYPRUS	MADAGASCAR	UKRAINIAN SOVIET SOCIALIST REPUBLIC
CZECHOSLOVAKIA	MALAYSIA	
DEMOCRATIC KAMPUCHEA	MALI	UNION OF SOVIET SOCIALIST REPUBLICS
DEMOCRATIC PEOPLE'S REPUBLIC OF KOREA	MAURITIUS	
	MEXICO	UNITED ARAB EMIRATES
DENMARK	MONACO	UNITED KINGDOM OF GREAT BRITAIN AND NORTHERN IRELAND
DOMINICAN REPUBLIC	MONGOLIA	
ECUADOR	MOROCCO	
EGYPT	MYANMAR	UNITED REPUBLIC OF TANZANIA
EL SALVADOR	NAMIBIA	
ETHIOPIA	NETHERLANDS	UNITED STATES OF AMERICA
FINLAND	NEW ZEALAND	URUGUAY
FRANCE	NICARAGUA	VENEZUELA
GABON	NIGER	VIET NAM
GERMANY	NIGERIA	YUGOSLAVIA
GHANA	NORWAY	ZAIRE
GREECE	PAKISTAN	ZAMBIA
GUATEMALA	PANAMA	ZIMBABWE
HAITI	PARAGUAY	

The Agency's Statute was approved on 23 October 1956 by the Conference on the Statute of the IAEA held at United Nations Headquarters, New York; it entered into force on 29 July 1957. The Headquarters of the Agency are situated in Vienna. Its principal objective is "to accelerate and enlarge the contribution of atomic energy to peace, health and prosperity throughout the world".

© IAEA, 1990

Permission to reproduce or translate the information contained in this publication may be obtained by writing to the International Atomic Energy Agency, Wagramerstrasse 5, P.O. Box 100, A-1400 Vienna, Austria.

Printed by the IAEA in Austria
November 1990

SAFETY SERIES No. 6

REGULATIONS
FOR THE SAFE TRANSPORT
OF RADIOACTIVE MATERIAL

1985 Edition
(As Amended 1990)

INTERNATIONAL ATOMIC ENERGY AGENCY
VIENNA, 1990

THESE REGULATIONS ARE ALSO PUBLISHED
IN FRENCH, RUSSIAN AND SPANISH

REGULATIONS FOR THE SAFE TRANSPORT
OF RADIOACTIVE MATERIAL, 1985 EDITION
(AS AMENDED 1990)
IAEA, VIENNA, 1990
STI/PUB/866
ISBN 92-0-123890-8
ISSN 0074-1892

FOREWORD

This publication is an updated version of the 1985 Edition of the Transport Regulations and replaces all previous publications of IAEA Safety Series No. 6. It includes the Supplements 1986 and 1988 to the Regulations, the minor changes adopted by the Review Panel meeting which convened in Vienna, 10–14 July 1989, and also the changes of detail which survived the 'ninety day rule' procedure which authorizes the Director General of the IAEA to promulgate such changes after giving Member States not less than ninety days notice and taking into account any comments they make.

Since this publication is an updated version of the 1985 Edition of the Transport Regulations, the old IAEA Safety Series style is maintained for the convenience of the user, although the old style has now generally been superseded by a new one. It should be noted that subsequent editions of the Regulations will be published in the new style.

CONTENTS

(Paragraph numbers are given in parentheses)

SECTION I. INTRODUCTION	1
Purpose and scope (101–109)	1
Definitions for the purpose of these Regulations (110–152)	2
SECTION II. GENERAL PROVISIONS	13
Radiation protection (201–206)	13
Emergency response (207–208)	14
Quality assurance (209)	15
Compliance assurance (210)	15
Special arrangement (211)	15
SECTION III. ACTIVITY AND FISSILE MATERIAL LIMITS	17
Basic A_1/A_2 values (301)	17
Determination of A_1 and A_2 (302–306)	17
Contents limits for packages (307–315)	31
SECTION IV. PREPARATION, REQUIREMENTS AND CONTROLS FOR SHIPMENT AND FOR STORAGE IN TRANSIT	33
Package inspection requirements (401–402)	33
Transport of other goods (403–406)	34
Other dangerous properties of contents (407)	34
Requirements and controls for contamination and for leaking packages (408–414)	34
Requirements and controls for transport of excepted packages (415–421)	36
Requirements and controls for transport of LSA material and SCO in industrial packages or unpackaged (422–427)	38
Determination of transport index (TI) (428–431)	40
Limits on transport index and radiation level for packages and overpacks (432–434)	43
Categories (435)	43

Marking, labelling and placarding (436–445) 45
Consignor's responsibilities (446–459) 51
Transport (460–477) 54
Storage in transit (478–482) 59
Customs operations (483) 59
Undeliverable packages (484) 59

SECTION V. REQUIREMENTS FOR RADIOACTIVE MATERIALS
AND FOR PACKAGINGS AND PACKAGES 61

Requirements for radioactive materials (501–504) 61
General requirements for all packagings and packages (505–514) 61
Additional requirements for packages transported by air (515–517) 62
Requirements for industrial packages (518–523) 63
Requirements for Type A packages (524–540) 64
Requirements for Type B packages (541–558) 66
Requirements for packages containing fissile material (559–568) 69

SECTION VI. TEST PROCEDURES 75

Demonstration of compliance (601–602) 75
Test for LSA-III material (603) 75
Tests for special form radioactive material (604–613) 76
Tests for packages (614–633) 78

SECTION VII. APPROVAL AND ADMINISTRATIVE
REQUIREMENTS 83

General (701) 83
Approval of special form radioactive material (702–703) 83
Approval of package designs (704–714) 84
Notification and registration of serial numbers (715) 86
Approval of shipments (716–719) 86
Approval of shipment under special arrangement (720–722) 87
Competent authority approval certificates (723–725) 87
Contents of approval certificates (726–729) 90
Validation of certificates (730) 94

APPENDIX I. EXCERPTS FROM LIST OF UNITED NATIONS
NUMBERS, PROPER SHIPPING NAME AND
DESCRIPTION AND SUBSIDIARY RISKS 95

APPENDIX II. CONVERSION FACTORS AND PREFIXES 97

CONTRIBUTORS TO DRAFTING AND REVIEW 99
SELECTION OF IAEA PUBLICATIONS RELATING TO THE
SAFE TRANSPORT OF RADIOACTIVE MATERIAL 105
INDEX 107

LIST OF TABLES

Table I	A_1 and A_2 values for radionuclides	18
Table II	General values for A_1 and A_2	31
Table III	Limits of non-fixed contamination on surfaces	35
Table IV	Activity limits for excepted packages	37
Table V	Industrial package requirements for LSA material and SCO	39
Table VI	Conveyance activity limits for LSA material and SCO in industrial packages or unpackaged	40
Table VII	Multiplication factors for large dimension loads	41
Table VIII	Determination of transport index	42
Table IX	Categories of packages	44
Table X	Categories of overpacks including freight containers when used as overpacks	44
Table XI	TI limits for freight containers and conveyances	56
Table XII	Insolation data	67
Table XIII	Limitations on homogeneous hydrogenous solutions or mixtures of fissile material	70
Table XIV	Free drop distance for testing packages to normal conditions of transport	79

Section I

INTRODUCTION

PURPOSE AND SCOPE

101. The purpose of these Regulations is to establish standards of safety which provide an acceptable level of control of the radiation hazards to persons, property and the environment that are associated with the transport of **radioactive material**. Controls instituted for other reasons, such as economics or physical protection, shall not detract from the standards of safety which these Regulations are intended to provide.

102. These Regulations shall apply to the transport of **radioactive material** other than that which is an integral part of the means of transport, by all modes on land, water, or in the air, including transport which is incidental to the use of the **radioactive materials**.

103. Transport shall be deemed to comprise all operations and conditions associated with and involved in the movement of **radioactive material**; these include the design, fabrication and maintenance of **packaging**, and the preparation, consigning, handling, carriage, storage in transit and receipt at the final destination of **packages**. Transport includes normal and accident conditions encountered in carriage and in storage during transit.

104. These Regulations do not apply:

(a) within establishments where the **radioactive material** is produced, used, or stored other than in the course of transport, and in respect of which other appropriate safety regulations are in force, or
(b) to human beings who have been implanted with radioisotopic cardiac pacemakers or other devices, or who have been treated with radiopharmaceuticals.

105. For **radioactive material** having other dangerous properties, and for transport or storage of **radioactive material** with other dangerous goods, the relevant transport regulations for dangerous goods of each of the countries through or into which the material is to be transported, and the regulations of the cognizant transport organizations, shall apply, in addition to these Regulations. It is also necessary to take into account the possible formation of products having dangerous properties by interaction of contents with the atmosphere or with water (e.g. the case of UF_6). (See paras 208 and 407.)

106. Taking into account the present levels of safety in the transport of **radioactive material**, it is not generally necessary to recommend routing restrictions. However, when such requirements are imposed, account shall be taken of all risks including normal and accident risks, both radiological and non-radiological.

107. In the transport of **radioactive material**, public and worker safety is assured when these Regulations are complied with. Confidence in this regard is achieved through **quality assurance** and **compliance assurance** programmes. **Quality assurance** involves plans and actions by designers and manufacturers of **packagings**, and by **consignors, carriers** and **competent authorities** to ensure that all requirements applicable to **packages** and **consignment** are properly met. **Compliance assurance** involves reviews, inspections and other actions aimed at confirming that the provisions of these Regulations are met in practice.

108. In certain parts of these Regulations, a particular action is prescribed, but the responsibility for carrying out the action is not specifically assigned to any particular person. Such responsibility may vary according to the laws and customs of different countries and the international conventions into which these countries have entered. For the purpose of these Regulations, it is not necessary to make this assignment, but only to identify the action itself. It remains the prerogative of each government to assign this responsibility.

109. In implementing the provisions of these Regulations, it may be necessary for Member States to issue complementary national regulations. Except as necessary for solely domestic purposes, such national regulations should not conflict with these Regulations.

DEFINITIONS FOR THE PURPOSE OF THESE REGULATIONS

A_1 and A_2

110. A_1 shall mean the maximum activity of **special form radioactive material** permitted in a **Type A package**. A_2 shall mean the maximum activity of **radioactive material**, other than **special form radioactive material**, permitted in a **Type A package**.

Aircraft

111. **Cargo aircraft** shall mean any aircraft, other than a **passenger aircraft**, which is carrying goods or property.

112. **Passenger aircraft** shall mean an aircraft that carries any person other than a crew member, a **carrier's** employee in an official capacity, an authorized representative of an appropriate national authority, or a person accompanying a **consignment**.

Approval

113. Multilateral approval shall mean approval by the relevant **competent authority** both of the country of origin of the **design** or **shipment** and of each country through or into which the **consignment** is to be transported. The term "through or into" specifically excludes "over", i.e. the approval and notification requirements shall not apply to a country over which **radioactive material** is carried in an **aircraft**, provided that there is no scheduled stop in that country.

114. Unilateral approval shall mean an approval of a **design** which is required to be given by the **competent authority** of the country of origin of the **design** only.

Carrier

115. Carrier shall mean any individual, organization or government undertaking the carriage of **radioactive material** by any means of transport. The term includes both carriers for hire or reward (known as common or contract carriers in some countries) and carriers on own account (known as private carriers in some countries).

Competent authority

116. Competent authority shall mean any national or international authority designated or otherwise recognized as such for any purpose in connection with these Regulations.

Compliance assurance

117. Compliance assurance shall mean a systematic programme of measures applied by a **competent authority** which is aimed at ensuring that the provisions of these Regulations are met in practice.

Consignee

118. Consignee shall mean any individual, organization or government which receives a **consignment**.

Consignment

119. Consignment shall mean any **package** or **packages**, or load of **radioactive material**, presented by a **consignor** for transport.

Consignor

120. **Consignor** shall mean any individual, organization or government which presents a **consignment** for transport, and is named as consignor in the transport documents.

Containment system

121. **Containment system** shall mean the assembly of components of the **packaging** specified by the designer as intended to retain the **radioactive material** during transport.

Contamination

122. **Contamination** shall mean the presence of a radioactive substance on a surface in quantities in excess of 0.4 Bq/cm^2 (10^{-5} μCi/cm^2) for beta and gamma emitters and low toxicity alpha emitters or 0.04 Bq/cm^2 (10^{-6} μCi/cm^2) for all other alpha emitters. Low toxicity alpha emitters are: **natural uranium; depleted uranium;** natural thorium; uranium-235 or uranium-238; thorium-232; thorium-228 and thorium-230 when contained in ores or physical or chemical concentrates; or alpha emitters with a half-life of less than 10 days.

123. **Fixed contamination** shall mean **contamination** other than **non-fixed contamination**.

124. **Non-fixed contamination** shall mean **contamination** that can be removed from a surface during normal handling.

Conveyance

125. **Conveyance** shall mean

(a) for transport by road or rail: any **vehicle**,
(b) for transport by water: any **vessel**, or any hold, compartment, or **defined deck area** of a **vessel**, and
(c) for transport by air: any **aircraft**.

Defined deck area

126. **Defined deck area** shall mean the area, of the weather deck of a **vessel**, or of a **vehicle** deck of a roll-on/roll-off ship or a ferry, which is allocated for the stowage of **radioactive material**.

Design

127. **Design** shall mean the description of **special form radioactive material**, **package** or **packaging** which enables such an item to be fully identified. The description may include specifications, engineering drawings, reports demonstrating compliance with regulatory requirements, and other relevant documentation.

Exclusive use

128. **Exclusive use** shall mean the sole use, by a single **consignor**, of a **conveyance** or of a large **freight container** with a minimum length of 6 m, in respect of which all initial, intermediate, and final loading and unloading is carried out in accordance with the directions of the **consignor** or **consignee**.

Fissile material

129. **Fissile material** shall mean uranium-233, uranium-235, plutonium-238, plutonium-239, plutonium-241, or any combination of these radionuclides. Unirradiated **natural uranium** and **depleted uranium,** and **natural uranium** or **depleted uranium** which has been irradiated in thermal reactors only, are not included in this definition.

Freight container

130. **Freight container** shall mean an article of transport equipment designed to facilitate the carriage of goods, either packaged or unpackaged, by one or more modes of transport without intermediate reloading. It shall be of a permanent enclosed character, rigid and strong enough for repeated use and must be fitted with devices facilitating its handling, particularly in transfer between **conveyances** and from one mode of transport to another. A small **freight container** is that which has either any overall outer dimension less than 1.5 m, or an internal volume of not more than 3.0 m^3. Any other **freight container** is considered to be a large **freight container**. A **freight container** may be used as a **packaging** if the applicable requirements are met. It may also be used to perform the function of an **overpack**.

Low specific activity material

131. **Low specific activity (LSA) material** shall mean **radioactive material** which by its nature has a limited **specific activity**, or **radioactive material** for which limits of estimated average **specific activity** apply. External shielding materials surrounding the **LSA material** shall not be considered in determining the estimated average **specific activity**.

LSA material shall be in one of three groups:

(a) **LSA-I**

    (i) Ores containing naturally occurring radionuclides (e.g. uranium, thorium), and uranium or thorium concentrates of such ores;

    (ii) Solid unirradiated **natural uranium** or **depleted uranium** or natural thorium or their solid or liquid compounds or mixtures; or

    (iii) **radioactive material**, other than **fissile material**, for which the A_2 value is unlimited.

(b) **LSA-II**

    (i) Water with tritium concentration up to 0.8 TBq/L (20 Ci/L); or

    (ii) Other material in which the activity is distributed throughout and the estimated average **specific activity** does not exceed 10^{-4} A_2/g for solids and gases, and 10^{-5} A_2/g for liquids.

(c) **LSA-III**

Solids (e.g. consolidated wastes, activated materials) in which:

    (i) The **radioactive material** is distributed throughout a solid or a collection of solid objects, or is essentially uniformly distributed in a solid compact binding agent (such as concrete, bitumen, ceramic, etc.);

    (ii) The **radioactive material** is relatively insoluble, or it is intrinsically contained in a relativy insoluble matrix, so that, even under loss of **packaging**, the loss of **radioactive material** per **package** by leaching when placed in water for seven days would not exceed 0.1 A_2; and

    (iii) The estimated average **specific activity** of the solid, excluding any shielding material, does not exceed 2×10^{-3} A_2/g.

Maximum normal operating pressure

132. **Maximum normal operating pressure** shall mean the maximum pressure above atmospheric pressure at mean sea level that would develop in the **containment system** in a period of one year under the conditions of temperature and solar radiation corresponding to environmental conditions of transport in the absence of venting, external cooling by an ancillary system, or operational controls during transport.

Overpack

133. **Overpack** shall mean an enclosure, such as a box or bag, which need not meet the requirements for a **freight container** and which is used by a single **consignor** to consolidate into one handling unit a **consignment** of two or more **packages** for convenience of handling, stowage and carriage.

Package

134. **Package** shall mean the **packaging** with its **radioactive contents** as presented for transport. **Package** and **packaging** performance standards, in terms of retention of integrity of containment and shielding, depend upon the quantity and nature of the **radioactive material** transported. Performance standards applied are graded to take into account conditions of transport characterized by the following severity levels:

— conditions likely to be encountered in routine transport (in incident free conditions),
— normal conditions of transport (minor mishaps), and
— accident conditions of transport.

The performance standards include design requirements and tests. Each **package** shall be classified as follows:

(a) **Excepted package** is a **packaging** containing **radioactive material** (see paras 418–420) that is designed to meet the General Requirements for All Packagings and Packages (see paras 505–514).

(b) (I) **Industrial package Type 1 (IP-1)** is a **packaging, tank** or **freight container** containing **LSA material** or **surface contaminated object (SCO)** (see paras 131, 144 and 426) that is designed to meet the General Requirements for All Packagings and Packages (see paras 505–514) and the requirements of paras 515–517 if carried by air;

(II) **Industrial package Type 2 (IP-2)** is a **packaging, tank** or **freight container** containing **LSA material** or **SCO** (see paras 131, 144 and 426), that is designed to meet the General Requirements for All Packagings and Packages (see paras 505–514), the requirements of paras 515–517 if carried by air, and, in addition, the following Specific Design Requirements:
 (i) for a **package**, see para. 519,
 (ii) for a **tank**, see paras 521–522, and
 (iii) for a **freight container**, see para. 523;

(III) **Industrial package Type 3 (IP-3)** is a **packaging, tank** or **freight container** containing **LSA material** or **SCO** (see paras 131, 144 and 426), that is designed to meet the General Requirements for All Packagings and Packages (see paras 505–514), the requirements of paras 515–517 if carried by air, and, in addition, the following Specific Design Requirements:
 (i) for a **package**, see para. 520,
 (ii) for a **tank**, see paras 521–522, and
 (iii) for a **freight container**, see para. 523;

(c) **Type A package** is a **packaging, tank** or **freight container** containing an activity up to A_1 if **special form radioactive material**, or up to A_2 if not **special form radioactive material**, that is designed to meet the General Requirements for All Packagings and Packages (see paras 505–514), the requirements of paras 515–517 if carried by air, and the Specific Design Requirements in paras 524–540, as appropriate;

(d) **Type B package** is a **packaging, tank** or **freight container** containing an activity that may be in excess of A_1, if **special form radioactive material,** or in excess of A_2 if not **special form radioactive material**, that is designed to meet the General Requirements for All Packagings and Packages (see paras 505–514), the requirements of paras 515–517 if carried by air, and the Specific Design Requirements in paras 525–538 and 541-558, as appropriate.

Packaging

135. **Packaging** shall mean the assembly of components necessary to enclose the **radioactive contents** completely. It may, in particular, consist of one or more receptacles, absorbent materials, spacing structures, radiation shielding, service equipment for filling, emptying, venting and pressure relief, and devices for cooling, for absorbing mechanical shocks, for providing handling and tie-down capability, for thermal insulation, and service devices integral to the **package**. The **packaging** may be a box, drum, or similar receptacle, or may also be a **freight container**, or **tank** consistent with para. 134.

Quality assurance

136. **Quality assurance** shall mean a systematic programme of controls and inspections applied by any organization or body involved in the transport of **radioactive material** which is aimed at providing adequate confidence that the standard of safety prescribed in these Regulations is achieved in practice.

Radiation level

137. **Radiation level** shall mean the corresponding dose equivalent rate expressed in millisieverts (previously millirem) per hour. (Note: it is recognized that millisieverts or millirem are not the correct units that should apply to radiation exposures in all cases; nevertheless, these units are used exclusively in these Regulations for convenience.)

Radioactive contents

138. **Radioactive contents** shall mean the **radioactive material** together with any contaminated solids, liquids, and gases within the **packaging**.

Radioactive material

139. **Radioactive material** shall mean any material having a **specific activity** greater than 70 kBq/kg (2 nCi/g).

Shipment

140. **Shipment** shall mean the specific movement of a **consignment** from origin to destination.

Special arrangement

141. **Special arrangement** shall mean those provisions, approved by the **competent authority**, under which a **consignment** which does not satisfy all the applicable requirements of these Regulations may be transported. For international **shipments** of this type **multilateral approval** is required. See para. 211.

Special form radioactive material

142. **Special form radioactive material** shall mean either an indispersible solid **radioactive material** or a sealed capsule containing **radioactive material**. See paras 502–504.

Specific activity

143. **Specific activity** shall mean the activity of a radionuclide per unit mass of that nuclide. The **specific activity** of a material in which the radionuclide is essentially uniformly distributed is the activity per unit mass of the material.

Surface contaminated object

144. **Surface contaminated object (SCO)** shall mean a solid object which is not itself radioactive but which has **radioactive material** distributed on its surfaces. SCO shall be in one of two groups:

(a) SCO-I: A solid object on which:

  (i) the **non-fixed contamination** on the accessible surface averaged over 300 cm^2 (or the area of the surface if less than 300 cm^2) does not exceed 4 Bq/cm^2 (10^{-4} μCi/cm^2) for beta and gamma emitters and low toxicity alpha emitters, or 0.4 Bq/cm^2 (10^{-5} μCi/cm^2) for all other alpha emitters; and

(ii) the **fixed contamination** on the accessible surface averaged over 300 cm² (or the area of the surface if less than 300 cm²) does not exceed 4×10^4 Bq/cm² (1 µCi/cm²) for beta and gamma emitters, and low toxicity alpha emitters, or 4×10^3 Bq/cm² (0.1 µCi/cm²) for all other alpha emitters; and

(iii) the **non-fixed contamination** plus the fixed contamination on the inaccessible surface averaged over 300 cm² (or the area of the surface if less than 300 cm²) does not exceed 4×10^4 Bq/cm² (1 µCi/cm²) for beta and gamma emitters and low toxicity alpha emitters, or 4×10^3 Bq/cm² (0.1 µCi/cm²) for all other alpha emitters.

(b) **SCO-II:** A solid object on which either the **fixed** or **non-fixed contamination** on the surface exceeds the applicable limits specified for **SCO-I** in (a) above and on which:

(i) the **non-fixed contamination** on the accessible surface averaged over 300 cm² (or the area of the surface if less than 300 cm²) does not exceed 400 Bq/cm² (10^{-2} µCi/cm²) for beta and gamma emitters and low toxicity alpha emitters, or 40 Bq/cm² (10^{-3} µCi/cm²) for all other alpha emitters; and

(ii) the **fixed contamination** on the accessible surface, averaged over 300 cm² (or the area of the surface if less than 300 cm²) does not exceed 8×10^5 Bq/cm² (20 µCi/cm²) for beta and gamma emitters and low toxicity alpha emitters, or 8×10^4 Bq/cm² (2 µCi/cm²) for all other alpha emitters; and

(iii) the **non-fixed contamination** plus the **fixed contamination** on the inaccessible surface averaged over 300 cm² (or the area of the surface if less than 300 cm²) does not exceed 8×10^5 Bq/cm² (20 µCi/cm²) for beta and gamma emitters and low toxicity alpha emitters, or 8×10^4 Bq/cm² (2 µCi/cm²) for all other alpha emitters.

Tank

145. **Tank** shall mean a tank container, portable tank, a road tank vehicle, a rail tank wagon or a receptacle with a capacity of not less than 450 litres to contain liquids, powders, granules, slurries or solids which are loaded as gas or liquid and subsequently solidified, and not less than 1000 litres to contain gases. A tank container shall be capable of being carried on land or on sea and of being loaded and discharged without the need of removal of its structural equipment, shall possess stabilizing members and tie-down attachments external to the shell, and shall be capable of being lifted when full.

Transport index

146. **Transport index (TI)** shall mean a single number assigned to a **package, overpack, tank** or **freight container**, or to unpackaged **LSA-I** or **SCO-I**, which is used to provide control over both nuclear criticality safety and radiation exposure. It is also used to establish contents limits on certain **packages, overpacks, tanks** and **freight containers**; to establish categories for labelling; to determine whether transport under **exclusive use** shall be required; to establish spacing requirements during storage in transit; to establish mixing restrictions during transport under **special arrangement** and during storage in transit; and to define the number of **packages** allowed in a **freight container** or aboard a **conveyance**. See Section IV.

Uncompressed gas

147. **Uncompressed gas** shall mean gas at a pressure not exceeding ambient atmospheric pressure at the time the **containment system** is closed.

Unirradiated thorium

148. **Unirradiated thorium** shall mean thorium containing not more than 10^{-7} g of uranium-233 per gram of thorium-232.

Unirradiated uranium

149. **Unirradiated uranium** shall mean uranium containing not more than 10^{-6} g of plutonium per gram of uranium-235 and not more than 9 MBq (0.20 mCi) of fission products per gram of uranium-235.

Uranium — natural, depleted, enriched

150. **Natural uranium** shall mean chemically separated uranium containing the naturally occurring distribution of uranium isotopes (approximately 99.28% uranium-238, and 0.72% uranium-235 by mass). **Depleted uranium** shall mean uranium containing a lesser mass percentage of uranium-235 than in **natural uranium**. **Enriched uranium** shall mean uranium containing a greater mass percentage of uranium-235 than in **natural uranium**. In all cases, a very small mass percentage of uranium-234 is present.

Vehicle

151. **Vehicle** shall mean a road vehicle (including an articulated vehicle, i.e. a tractor and semi-trailer combination) or railroad car or railway wagon. Each trailer shall be considered as a separate **vehicle**.

Vessel

152. **Vessel** shall mean any seagoing vessel or inland waterway craft used for carrying cargo.

Section II

GENERAL PROVISIONS

RADIATION PROTECTION

201. The radiation exposure of transport workers and of the general public is subject to the requirements specified in the "Basic Safety Standards for Radiation Protection, 1982 Edition", Safety Series No. 9, IAEA, Vienna (1982), jointly sponsored by the IAEA, ILO, NEA/OECD and WHO.

202. Radiation exposures from the handling, storage and transport of **radioactive material** shall be kept as low as reasonably achievable, economic and social factors being taken into account. Compliance with these Regulations and with the Basic Safety Standards for Radiation Protection will ensure a high degree of safety, but managers and workers have a continuous responsibility for maintaining safe working practices. Transport workers shall receive appropriate training (to the extent necessary, considering the type of work) concerning the radiation hazards involved and the precautions to be observed.

203. The relevant **competent authority** shall arrange for periodic assessments to be carried out as necessary to evaluate the radiation doses to workers and to members of the public due to the transport of **radioactive material**, to (1) ensure the implementation of operational requirements for keeping radiation exposures as low as reasonably achievable, and (2) ensure that the system of dose limitation for transport workers and members of the public, as set forth in the Agency's Basic Safety Standards for Radiation Protection, is being complied with.

204. The nature and extent of the measures to be employed in controlling radiation exposures shall be related to the magnitude and likelihood of the exposures. Administrative requirements applicable to transport workers are set forth in Section V of the Basic Safety Standards for Radiation Protection. For individual occupationally exposed workers, where it is determined that the dose received

(a) is most unlikely to exceed 5 mSv (500 mrem) per year, neither special work patterns nor detailed monitoring or assessment of radiation doses shall be required;
(b) is likely to be between 5 mSv (500 mrem) and 15 mSv (1500 mrem) per year, periodic (as necessary) environmental monitoring and assessments of radiation exposure levels in work areas (including in **conveyances**) shall be conducted; and
(c) is likely to be between 15 mSv (1500 mrem) and 50 mSv (5000 mrem) per year, individual radiation exposure monitoring programmes and special health supervision shall be required.

205. **Radioactive material** shall be segregated sufficiently from transport workers and from members of the public. Different limiting values for dose, only for the purposes of calculating segregation distances or dose rates in regularly occupied areas shall be required:

(a) For transport workers, in the determination of segregation distances or dose rates in regularly occupied working areas, a dose level of 5 mSv (500 mrem) per year shall be used as the limiting value. This value, together with hypothetical but realistic mathematical models and parameters, shall be used to determine segregation distances or associated dose rates for transport workers.

(b) For members of the public, in the determination of segregation distances or dose rates in regularly occupied public areas or in areas where the public has regular access, a dose level of not more than 1 mSv (100 mrem) per year to the critical group shall be used as the limiting value. This value shall be used together with hypothetical but realistic models and parameters to determine segregation distances or dose rates for members of the public, with the objective of providing reasonable assurance that actual doses from transport of **radioactive material** will not exceed small fractions of the appropriate dose limits.

206. **Radioactive material** shall be sufficiently segregated from undeveloped photographic film. The basis for determining segregation distances for this purpose shall be that the radiation exposure of undeveloped photographic film due to the transport of **radioactive material** be limited to 0.1 mSv (10 mrem) per **consignment** of such film.

EMERGENCY RESPONSE

207. In the event of accidents during the transport of **radioactive material**, emergency provisions, as established by relevant national and/or international organizations, shall be observed in order to protect human health and minimize danger to life and property. Appropriate guidelines for such provisions are contained in "Emergency Response Planning and Preparedness for Transport Accidents Involving Radioactive Material", Safety Series No. 87, IAEA, Vienna (1988).

208. Account shall be taken of the formation of other dangerous substances that may result from the reaction between the contents of a **consignment** and the atmosphere or water in the event of breaking of the **containment system** caused by an accident, e.g. UF_6 decomposition in a humid atmosphere.

QUALITY ASSURANCE

209. **Quality assurance** programmes shall be established for the design, manufacture, testing, documentation, use, maintenance and inspection of all **packages** and for transport and in-transit storage operations to ensure compliance with the relevant provisions of these Regulations. Where **competent authority approval** for **design** or **shipment** is required, such **approval** shall take into account and be contingent upon the adequacy of the **quality assurance** programme. Certification that the **design** specification has been fully implemented shall be available to the **competent authority**. The manufacturer, **consignor**, or user of any **package design** shall be prepared to provide facilities for **competent authority** inspection of the **packaging** during construction and use and to demonstrate to any cognizant **competent authority** that:

(a) the construction methods and materials used for the construction of the **packaging** are in accordance with the approved **design** specifications; and

(b) all **packagings** built to an approved **design** are periodically inspected and, as necessary, repaired and maintained in good condition so that they continue to comply with all relevant requirements and specifications, even after repeated use.

COMPLIANCE ASSURANCE

210. The **competent authority** is responsible for assuring compliance with these Regulations. Means to discharge this responsibility include the establishment and execution of a programme for monitoring the design, manufacture, testing, inspection and maintenance of **packaging**, and the preparation, documentation, handling and stowage of **packages** by **consignors** and **carriers** respectively, to provide evidence that the provisions of these Regulations are being met in practice.

SPECIAL ARRANGEMENT

211. A **consignment** which does not satisfy all the applicable requirements of these Regulations shall not be transported except under **special arrangement**. Provisions may be approved by a **competent authority**, under which a **consignment**, which does not satisfy all of the applicable requirements of these Regulations, may be transported under **special arrangement**. These provisions shall be adequate to ensure that the overall level of safety in transport and in-transit storage is at least equivalent to that which would be provided if all the applicable requirements had been met. For international **consignments** of this type, **multilateral approval** shall be required.

Section III

ACTIVITY AND FISSILE MATERIAL LIMITS

BASIC A_1/A_2 VALUES

301. Values of A_1 and A_2 for individual radionuclides, which are the bases for many activity limits elsewhere in these Regulations, are given in Table I.

DETERMINATION OF A_1 AND A_2

302. For individual radionuclides whose identities are known, but which are not listed in Table I, the determination of the values of A_1 and A_2 shall require **competent authority** approval or, for international transport, **multilateral approval**. Alternatively, the values of A_1 and A_2 in Table II may be used without obtaining **competent authority** approval.

303. In the calculations of A_1 and A_2 for a radionuclide not in Table I, a single radioactive decay chain in which the radionuclides are present in their naturally occurring proportions and in which no daughter nuclide has a half-life either longer than 10 days or longer than that of the parent nuclide shall be considered as a single radionuclide, and the activity to be taken into account and the A_1 or A_2 value to be applied shall be those corresponding to the parent nuclide of that chain. In the case of radioactive decay chains in which any daughter nuclide has a half-life either longer than 10 days or greater than that of the parent nuclide, the parent and such daughter nuclides shall be considered as mixtures of different nuclides.

304. For mixtures of radionuclides whose identities and respective activities are known the following conditions shall apply:

(a) For **special form radioactive material**:

$$\sum_i \frac{B(i)}{A_1(i)} \text{ less than or equal to 1}$$

(b) For other forms of **radioactive material**:

$$\sum_i \frac{B(i)}{A_2(i)} \text{ less than or equal to 1}$$

where B(i) is the activity of radionuclide i and $A_1(i)$ and $A_2(i)$ are the A_1 and A_2 values for radionuclide i, respectively.

TABLE I. A_1 AND A_2 VALUES FOR RADIONUCLIDES

Symbol of radionuclide	Element and atomic number	A_1 (TBq)	A_1 (Ci) (approx.[a])	A_2 (TBq)	A_2 (Ci) (approx.[a])
^{225}Ac (b)*	Actinium (89)	0.6	10	1×10^{-2}	2×10^{-1}
^{227}Ac		40	1000	2×10^{-5}	5×10^{-4}
^{228}Ac		0.6	10	0.4	10
^{105}Ag	Silver (47)	2	50	2	50
^{108}Ag^m		0.6	10	0.6	10
^{110}Ag^m		0.4	10	0.4	10
^{111}Ag		0.6	10	0.5	10
^{26}Al	Aluminium (13)	0.4	10	0.4	10
^{241}Am	Americium (95)	2	50	2×10^{-4}	5×10^{-3}
^{242}Am^m		2	50	2×10^{-4}	5×10^{-3}
^{243}Am		2	50	2×10^{-4}	5×10^{-3}
^{37}Ar	Argon (18)	40	1000	40	1000
^{39}Ar		20	500	20	500
^{41}Ar		0.6	10	0.6	10
^{42}Ar (b)		0.2	5	0.2	5
^{72}As	Arsenic (33)	0.2	5	0.2	5
^{73}As		40	1000	40	1000
^{74}As		1	20	0.5	10
^{76}As		0.2	5	0.2	5
^{77}As		20	500	0.5	10
^{211}At	Astatine (85)	30	800	2	50
^{193}Au	Gold (79)	6	100	6	100
^{194}Au		1	20	1	20
^{195}Au		10	200	10	200
^{196}Au		2	50	2	50
^{198}Au		3	80	0.5	10
^{199}Au		10	200	0.9	20

* Note: (b) indicates a footnote at the end of Table I: this form is used here to avoid confusion with the superscript m.

TABLE I. (cont.)

Symbol of radionuclide	Element and atomic number	A_1 (TBq)	A_1 (Ci) (approx.[a])	A_2 (TBq)	A_2 (Ci) (approx.[a])
^{131}Ba	Barium (56)	2	50	2	50
^{133}Ba^m		10	200	0.9	20
^{133}Ba		3	80	3	80
^{140}Ba (b)		0.4	10	0.4	10
^{7}Be	Beryllium (4)	20	500	20	500
^{10}Be		20	500	0.5	10
^{205}Bi	Bismuth (83)	0.6	10	0.6	10
^{206}Bi		0.3	8	0.3	8
^{207}Bi		0.7	10	0.7	10
^{210}Bi^m (b)		0.3	8	3×10^{-2}	8×10^{-1}
^{210}Bi		0.6	10	0.5	10
^{212}Bi (b)		0.3	8	0.3	8
^{247}Bk	Berkelium (97)	2	50	2×10^{-4}	5×10^{-3}
^{249}Bk		40	1000	8×10^{-2}	2
^{76}Br	Bromine (35)	0.3	8	0.3	8
^{77}Br		3	80	3	80
^{82}Br		0.4	10	0.4	10
^{11}C	Carbon (6)	1	20	0.5	10
^{14}C		40	1000	2	50
^{41}Ca	Calcium (20)	40	1000	40	1000
^{45}Ca		40	1000	0.9	20
^{47}Ca		0.9	20	0.5	10
^{109}Cd	Cadmium (48)	40	1000	1	20
^{113}Cd^m		20	500	9×10^{-2}	2
^{115}Cd^m		0.3	8	0.3	8
^{115}Cd		4	100	0.5	10
^{139}Ce	Cerium (58)	6	100	6	100
^{141}Ce		10	200	0.5	10
^{143}Ce		0.6	10	0.5	10
^{144}Ce (b)		0.2	5	0.2	5

For footnotes see page 30.

TABLE I. (cont.)

Symbol of radionuclide	Element and atomic number	A_1 (TBq)	A_1 (Ci) (approx.[a])	A_2 (TBq)	A_2 (Ci) (approx.[a])
^{248}Cf	Californium (98)	30	800	3×10^{-3}	8×10^{-2}
^{249}Cf		2	50	2×10^{-4}	5×10^{-3}
^{250}Cf		5	100	5×10^{-4}	1×10^{-2}
^{251}Cf		2	50	2×10^{-4}	5×10^{-3}
^{252}Cf		0.1	2	1×10^{-3}	2×10^{-2}
^{253}Cf		40	1000	6×10^{-2}	1
^{254}Cf		3×10^{-3}	8×10^{-2}	6×10^{-4}	1×10^{-2}
^{36}Cl	Chlorine (17)	20	500	0.5	10
^{38}Cl		0.2	5	0.2	5
^{240}Cm	Curium (96)	40	1000	2×10^{-2}	5×10^{-1}
^{241}Cm		2	50	0.9	20
^{242}Cm		40	1000	1×10^{-2}	2×10^{-1}
^{243}Cm		3	80	3×10^{-4}	8×10^{-3}
^{244}Cm		4	100	4×10^{-4}	1×10^{-2}
^{245}Cm		2	50	2×10^{-4}	5×10^{-3}
^{246}Cm		2	50	2×10^{-4}	5×10^{-3}
^{247}Cm		2	50	2×10^{-4}	5×10^{-3}
^{248}Cm		4×10^{-2}	1	5×10^{-5}	1×10^{-3}
^{55}Co	Cobalt (27)	0.5	10	0.5	10
^{56}Co		0.3	8	0.3	8
^{57}Co		8	200	8	200
^{58}Co^m		40	1000	40	1000
^{58}Co		1	20	1	20
^{60}Co		0.4	10	0.4	10
^{51}Cr	Chromium (24)	30	800	30	800
^{129}Cs	Caesium (55)	4	100	4	100
^{131}Cs		40	1000	40	1000
^{132}Cs		1	20	1	20
^{134}Cs^m		40	1000	9	200
^{134}Cs		0.6	10	0.5	10

For footnotes see page 30.

TABLE I. (cont.)

Symbol of radionuclide	Element and atomic number	A_1 (TBq)	A_1 (Ci) (approx.[a])	A_2 (TBq)	A_2 (Ci) (approx.[a])
^{135}Cs		40	1000	0.9	20
^{136}Cs		0.5	10	0.5	10
^{137}Cs [b]		2	50	0.5	10
^{64}Cu	Copper (29)	5	100	0.9	20
^{67}Cu		9	200	0.9	20
^{159}Dy	Dysprosium (66)	20	500	20	500
^{165}Dy		0.6	10	0.5	10
^{166}Dy [b]		0.3	8	0.3	8
^{169}Er	Erbium (68)	40	1000	0.9	20
^{171}Er		0.6	10	0.5	10
^{147}Eu	Europium (63)	2	50	2	50
^{148}Eu		0.5	10	0.5	10
^{149}Eu		20	500	20	500
^{150}Eu		0.7	10	0.7	10
^{152}Eu^m		0.6	10	0.5	10
^{152}Eu		0.9	20	0.9	20
^{154}Eu		0.8	20	0.5	10
^{155}Eu		20	500	2	50
^{156}Eu		0.6	10	0.5	10
^{18}F	Fluorine (9)	1	20	0.5	10
^{52}Fe [b]	Iron (26)	0.2	5	0.2	5
^{55}Fe		40	1000	40	1000
^{59}Fe		0.8	20	0.8	20
^{60}Fe		40	1000	0.2	5
^{67}Ga	Gallium (31)	6	100	6	100
^{68}Ga		0.3	8	0.3	8
^{72}Ga		0.4	10	0.4	10
^{146}Gd [b]	Gadolinium (64)	0.4	10	0.4	10
^{148}Gd		3	80	3×10^{-4}	8×10^{-3}
^{153}Gd		10	200	5	100
^{159}Gd		4	100	0.5	10

For footnotes see page 30.

TABLE I. (cont.)

Symbol of radionuclide	Element and atomic number	A_1 (TBq)	A_1 (Ci) (approx.[a])	A_2 (TBq)	A_2 (Ci) (approx.[a])
^{68}Ge (b)	Germanium (32)	0.3	8	0.3	8
^{71}Ge		40	1000	40	1000
^{77}Ge		0.3	8	0.3	8
^{172}Hf(b)	Hafnium (72)	0.5	10	0.3	8
^{175}Hf		3	80	3	80
^{181}Hf		2	50	0.9	20
^{182}Hf		4	100	3×10^{-2}	8×10^{-1}
^{194}Hg (b)	Mercury (80)	1	20	1	20
^{195}Hg^m		5	100	5	100
^{197}Hg^m		10	200	0.9	20
^{197}Hg		10	200	10	200
^{203}Hg		4	100	0.9	20
^{163}Ho	Holmium (67)	40	1000	40	1000
^{166}Ho^m		0.6	10	0.3	8
^{166}Ho		0.3	8	0.3	8
^{123}I	Iodine (53)	6	100	6	100
^{124}I		0.9	20	0.9	20
^{125}I		20	500	2	50
^{126}I		2	50	0.9	20
^{129}I		Unlimited		Unlimited	
^{131}I		3	80	0.5	10
^{132}I		0.4	10	0.4	10
^{133}I		0.6	10	0.5	10
^{134}I		0.3	8	0.3	8
^{135}I		0.6	10	0.5	10
^{111}In	Indium (49)	2	50	2	50
^{113}In^m		4	100	4	100
^{114}In^m (b)		0.3	8	0.3	8
^{115}In^m		6	100	0.9	20

For footnotes see page 30.

TABLE I. (cont.)

Symbol of radionuclide	Element and atomic number	A_1 (TBq)	A_1 (Ci) (approx.[a])	A_2 (TBq)	A_2 (Ci) (approx.[a])
^{189}Ir	Iridium (77)	10	200	10	200
^{190}Ir		0.7	10	0.7	10
^{192}Ir		1	20	0.5	10
^{193}Ir^m		10	200	10	200
^{194}Ir		0.2	5	0.2	5
^{40}K	Potassium (19)	0.6	10	0.6	10
^{42}K		0.2	5	0.2	5
^{43}K		1	20	0.5	10
^{81}Kr	Krypton (36)	40	1000	40	1000
^{85}Kr^m		6	100	6	100
^{85}Kr		20	500	10	200
^{87}Kr		0.2	5	0.2	5
^{137}La	Lanthanum (57)	40	1000	2	50
^{140}La		0.4	10	0.4	10
LSA	Low specific activity material (see para. 131)				
^{172}Lu	Lutetium (71)	0.5	10	0.5	10
^{173}Lu		8	200	8	200
^{174}Lu^m		20	500	8	200
^{174}Lu		8	200	4	100
^{177}Lu		30	800	0.9	20
MFP	For mixed fission products, use formula for mixtures or Table II				
^{28}Mg(b)	Magnesium (12)	0.2	5	0.2	5
^{52}Mn	Manganese (25)	0.3	8	0.3	8
^{53}Mn		Unlimited		Unlimited	
^{54}Mn		1	20	1	20
^{56}Mn		0.2	5	0.2	5
^{93}Mo	Molybdenum (42)	40	1000	7	100
^{99}Mo		0.6	10	0.5	10
^{13}N	Nitrogen (7)	0.6	10	0.5	10
^{22}Na	Sodium (11)	0.5	10	0.5	10
^{24}Na		0.2	5	0.2	5

For footnotes see page 30.

TABLE I. (cont.)

Symbol of radionuclide	Element and atomic number	A_1 (TBq)	A_1 (Ci) (approx.[a])	A_2 (TBq)	A_2 (Ci) (approx.[a])
^{92}Nb^m	Niobium (41)	0.7	10	0.7	10
^{93}Nb^m		40	1000	6	100
^{94}Nb		0.6	10	0.6	10
^{95}Nb		1	20	1	20
^{97}Nb		0.6	10	0.5	10
^{147}Nd	Neodymium (60)	4	100	0.5	10
^{149}Nd		0.6	10	0.5	10
^{59}Ni	Nickel (28)	40	1000	40	1000
^{63}Ni		40	1000	30	800
^{65}Ni		0.3	8	0.3	8
^{235}Np	Neptunium (93)	40	1000	40	1000
^{236}Np		7	100	1×10^{-3}	2×10^{-2}
^{237}Np		2	50	2×10^{-4}	5×10^{-3}
^{239}Np		6	100	0.5	10
^{185}Os	Osmium (76)	1	20	1	20
^{191}Os^m		40	1000	40	1000
^{191}Os		10	200	0.9	20
^{193}Os		0.6	10	0.5	10
^{194}Os (b)		0.2	5	0.2	5
^{32}P	Phosphorus (15)	0.3	8	0.3	8
^{33}P		40	1000	0.9	20
^{230}Pa	Protactinium (91)	2	50	0.1	2
^{231}Pa		0.6	10	6×10^{-5}	1×10^{-3}
^{233}Pa		5	100	0.9	20
^{201}Pb	Lead (82)	1	20	1	20
^{202}Pb		40	1000	2	50
^{203}Pb		3	80	3	80
^{205}Pb		Unlimited		Unlimited	
^{210}Pb (b)		0.6	10	9×10^{-3}	2×10^{-1}
^{212}Pb (b)		0.3	8	0.3	8
^{103}Pd	Palladium (46)	40	1000	40	1000
^{107}Pd		Unlimited		Unlimited	
^{109}Pd		0.6	10	0.5	10

For footnotes see page 30.

TABLE I. (cont.)

Symbol of radionuclide	Element and atomic number	A_1 (TBq)	A_1 (Ci) (approx.[a])	A_2 (TBq)	A_2 (Ci) (approx.[a])
^{143}Pm	Promethium (61)	3	80	3	80
^{144}Pm		0.6	10	0.6	10
^{145}Pm		30	800	7	100
^{147}Pm		40	1000	0.9	20
^{148}Pm^m		0.5	10	0.5	10
^{149}Pm		0.6	10	0.5	10
^{151}Pm		3	80	0.5	10
^{208}Po	Polonium (84)	40	1000	2×10^{-2}	5×10^{-1}
^{209}Po		40	1000	2×10^{-2}	5×10^{-1}
^{210}Po		40	1000	2×10^{-2}	5×10^{-1}
^{142}Pr	Praseodymium (59)	0.2	5	0.2	5
^{143}Pr		4	100	0.5	10
^{188}Pt (b)	Platinum (78)	0.6	10	0.6	10
^{191}Pt		3	80	3	80
^{193}Pt^m		40	1000	9	200
^{193}Pt		40	1000	40	1000
^{195}Pt^m		10	200	2	50
^{197}Pt^m		10	200	0.9	20
^{197}Pt		20	500	0.5	10
^{236}Pu	Plutonium (94)	7	100	7×10^{-4}	1×10^{-2}
^{237}Pu		20	500	20	500
^{238}Pu		2	50	2×10^{-4}	5×10^{-3}
^{239}Pu		2	50	2×10^{-4}	5×10^{-3}
^{240}Pu		2	50	2×10^{-4}	5×10^{-3}
^{241}Pu		40	1000	1×10^{-2}	2×10^{-1}
^{242}Pu		2	50	2×10^{-4}	5×10^{-3}
^{244}Pu (b)		0.3	8	2×10^{-4}	5×10^{-3}
^{223}Ra (b)	Radium (88)	0.6	10	3×10^{-2}	8×10^{-1}
^{224}Ra (b)		0.3	8	6×10^{-2}	1
^{225}Ra (b)		0.6	10	2×10^{-2}	5×10^{-1}
^{226}Ra (b)		0.3	8	2×10^{-2}	5×10^{-1}
^{228}Ra (b)		0.6	10	4×10^{-2}	1

For footnotes see page 30.

TABLE I. (cont.)

Symbol of radionuclide	Element and atomic number	A_1 (TBq)	A_1 (Ci) (approx.[a])	A_2 (TBq)	A_2 (Ci) (approx.[a])
^{81}Rb	Rubidium (37)	2	50	0.9	20
^{83}Rb		2	50	2	50
^{84}Rb		1	20	0.9	20
^{86}Rb		0.3	8	0.3	8
^{87}Rb		Unlimited		Unlimited	
Rb (natural)		Unlimited		Unlimited	
^{183}Re	Rhenium (75)	5	100	5	100
^{184}Re^m		3	80	3	80
^{184}Re		1	20	1	20
^{186}Re		4	100	0.5	10
^{187}Re		Unlimited		Unlimited	
^{188}Re		0.2	5	0.2	5
^{189}Re		4	100	0.5	10
Re (natural)		Unlimited		Unlimited	
^{99}Rh	Rhodium (45)	2	50	2	50
^{101}Rh		4	100	4	100
^{102}Rh^m		2	50	0.9	20
^{102}Rh		0.5	10	0.5	10
^{103}Rh^m		40	1000	40	1000
^{105}Rh		10	200	0.9	20
^{222}Rn (b)	Radon (86)	0.2	5	4×10^{-3}	1×10^{-1}
^{97}Ru	Ruthenium (44)	4	100	4	100
^{103}Ru		2	50	0.9	20
^{105}Ru		0.6	10	0.5	10
^{106}Ru (b)		0.2	5	0.2	5
^{35}S	Sulphur (16)	40	1000	2	50
^{122}Sb	Antimony (51)	0.3	8	0.3	8
^{124}Sb		0.6	10	0.5	10
^{125}Sb		2	50	0.9	20
^{126}Sb		0.4	10	0.4	10
^{44}Sc	Scandium (21)	0.5	10	0.5	10
^{46}Sc		0.5	10	0.5	10
^{47}Sc		9	200	0.9	20
^{48}Sc		0.3	8	0.3	8

For footnotes see page 30.

TABLE I. (cont.)

Symbol of radionuclide	Element and atomic number	A_1 (TBq)	A_1 (Ci) (approx.[a])	A_2 (TBq)	A_2 (Ci) (approx.[a])
SCO	Surface contaminated objects (see para. 144)				
^{75}Se	Selenium (34)	3	80	3	80
^{79}Se		40	1000	2	50
^{31}Si	Silicon (14)	0.6	10	0.5	10
^{32}Si		40	1000	0.2	5
^{145}Sm	Samarium (62)	20	500	20	500
^{147}Sm		Unlimited		Unlimited	
^{151}Sm		40	1000	4	100
^{153}Sm		4	100	0.5	10
^{113}Sn (b)	Tin (50)	4	100	4	100
^{117}Sn^m		6	100	2	50
^{119}Sn^m		40	1000	40	1000
^{121}Sn^m		40	1000	0.9	20
^{123}Sn		0.6	10	0.5	10
^{125}Sn		0.2	5	0.2	5
^{126}Sn (b)		0.3	8	0.3	8
^{82}Sr (b)	Strontium (38)	0.2	5	0.2	5
^{85}Sr^m		5	100	5	100
^{85}Sr		2	50	2	50
^{87}Sr^m		3	80	3	80
^{89}Sr		0.6	10	0.5	10
^{90}Sr (b)		0.2	5	0.1	2
^{91}Sr		0.3	8	0.3	8
^{92}Sr (b)		0.8	5	0.5	10
T (all forms)	Tritium (1)	40	1000	40	1000
^{178}Ta	Tantalum (73)	1	20	1	20
^{179}Ta		30	800	30	800
^{182}Ta		0.8	20	0.5	10
^{157}Tb	Terbium (65)	40	1000	10	200
^{158}Tb		1	20	0.7	10
^{160}Tb		0.9	20	0.5	10

For footnotes see page 30.

TABLE I. (cont.)

Symbol of radionuclide	Element and atomic number	A_1 (TBq)	A_1 (Ci) (approx.[a])	A_2 (TBq)	A_2 (Ci) (approx.[a])
$^{95}Tc^m$	Technetium (43)	2	50	2	50
$^{96}Tc^m$ (b)		0.4	10	0.4	10
^{96}Tc		0.4	10	0.4	10
$^{97}Tc^m$		40	1000	40	1000
^{97}Tc		Unlimited		Unlimited	
^{98}Tc		0.7	10	0.7	10
$^{99}Tc^m$		8	200	8	200
^{99}Tc		40	1000	0.9	20
^{118}Te (b)	Tellurium (52)	0.2	5	0.2	5
$^{121}Te^m$		5	100	5	100
^{121}Te		2	50	2	50
$^{123}Te^m$		7	100	7	100
$^{125}Te^m$		30	800	9	200
$^{127}Te^m$ (b)		20	500	0.5	10
^{127}Te		20	500	0.5	10
$^{129}Te^m$ (b)		0.6	10	0.5	10
^{129}Te		0.6	10	0.5	10
$^{131}Te^m$		0.7	10	0.5	10
^{132}Te (b)		0.4	10	0.4	10
^{227}Th	Thorium (90)	9	200	1×10^{-2}	2×10^{-1}
^{228}Th (b)		0.3	8	4×10^{-4}	1×10^{-2}
^{229}Th		0.3	8	3×10^{-5}	8×10^{-4}
^{230}Th		2	50	2×10^{-4}	5×10^{-3}
^{231}Th		40	1000	0.9	20
^{232}Th		Unlimited		Unlimited	
^{234}Th (b)		0.2	5	0.2	5
Th (natural)		Unlimited		Unlimited	
^{44}Ti (b)	Titanium (22)	0.5	10	0.2	5
^{200}Tl	Thallium (81)	0.8	20	0.8	20
^{201}Tl		10	200	10	200
^{202}Tl		2	50	2	50
^{204}Tl		4	100	0.5	10
^{167}Tm	Thulium (69)	7	100	7	100
^{168}Tm		0.8	20	0.8	20
^{170}Tm		4	100	0.5	10
^{171}Tm		40	1000	10	200

For footnotes see page 30.

TABLE I. (cont.)

Symbol of radionuclide	Element and atomic number	A_1 (TBq)	A_1 (Ci) (approx.[a])	A_2 (TBq)	A_2 (Ci) (approx.[a])
^{230}U	Uranium (92)	40	1000	1×10^{-2}	2×10^{-1}
^{232}U		3	80	3×10^{-4}	8×10^{-3}
^{233}U		10	200	1×10^{-3}	2×10^{-2}
^{234}U		10	200	1×10^{-3}	2×10^{-2}
^{235}U		Unlimited[c]		Unlimited[c]	
^{236}U		10	200	1×10^{-3}	2×10^{-2}
^{238}U		Unlimited		Unlimited	
U (natural)		Unlimited		Unlimited[d]	
U (enriched 5% or less)		Unlimited[c]		Unlimited[c,d]	
U (enriched more than 5%)		10	200	1×10^{-3} [d]	2×10^{-2}
U (depleted)		Unlimited		Unlimited[d]	
^{48}V	Vanadium (23)	0.3	8	0.3	8
^{49}V		40	1000	40	1000
^{178}W (b)	Tungsten (74)	1	20	1	20
^{181}W		30	800	30	800
^{185}W		40	1000	0.9	20
^{187}W		2	50	0.5	10
^{188}W (b)		0.2	5	0.2	5
^{122}Xe (b)	Xenon (54)	0.2	5	0.2	5
^{123}Xe		0.2	5	0.2	5
^{127}Xe		4	100	4	100
^{131}Xe^m		40	1000	40	1000
^{133}Xe		20	500	20	500
^{135}Xe		4	100	4	100
^{87}Y	Yttrium (39)	2	50	2	50
^{88}Y		0.4	10	0.4	10
^{90}Y		0.2	5	0.2	5
^{91}Y^m		2	50	2	50
^{91}Y		0.3	8	0.3	8
^{92}Y		0.2	5	0.2	5
^{93}Y		0.2	5	0.2	5
^{169}Yb	Ytterbium (70)	3	80	3	80
^{175}Yb		30	800	0.9	20

For footnotes see page 30.

TABLE I. (cont.)

Symbol of radionuclide	Element and atomic number	A_1 (TBq)	A_1 (Ci) (approx.[a])	A_2 (TBq)	A_2 (Ci) (approx.[a])
^{65}Zn	Zinc (30)	2	50	2	50
^{69}Zn^m (b)		2	50	0.5	10
^{69}Zn		4	100	0.5	10
^{88}Zr	Zirconium (40)	3	80	3	80
^{93}Zr		40	1000	0.2	5
^{95}Zr		1	20	0.9	20
^{97}Zr		0.3	8	0.3	8

[a] The curie values quoted are obtained by rounding down from the TBq figure after conversion to Ci. This ensures that the magnitude of A_1 or A_2 in Ci is always less than that in TBq.

[b] A_1 and/or A_2 value limited by daughter product decay.

[c] A_1 and A_2 are unlimited for radiation control purposes only. For nuclear criticality safety this material is subject to the control placed on **fissile material**.

[d] These values do not apply to reprocessed uranium.

Alternatively, an A_2 value for mixtures may be determined as follows:

$$A_2 \text{ for mixture} = \frac{1}{\sum_i \frac{f(i)}{A_2(i)}}$$

where f(i) is the fraction of activity of nuclide i in the mixture and A_2(i) is the appropriate A_2 value for nuclide i.

305. When the identity of each radionuclide is known but the individual activities of some of the radionuclides are not known, the radionuclides may be grouped and the lowest A_1 or A_2 value, as appropriate, for the radionuclides in each group may be used in applying the formulas in para. 304. Groups may be based on the total alpha activity and the total beta/gamma activity when these are known, using the lowest A_1 or A_2 values for the alpha emitters or beta/gamma emitters, respectively.

306. For individual radionuclides or for mixtures of radionuclides for which relevant data are not available, the values shown in Table II shall be used.

TABLE II. GENERAL VALUES FOR A_1 AND A_2

Contents	A_1		A_2	
	TBq	(Ci)[a]	TBq	(Ci)[a]
Only beta or gamma emitting nuclides are known to be present	0.2	(5)	0.02	(0.5)
Alpha emitting nuclides are known to be present or no relevant data are available	0.1	(2)	2×10^{-5}	(5×10^{-4})

[a] The curie values quoted in parentheses are approximate values and are not higher than the TBq values.

CONTENTS LIMITS FOR PACKAGES

307. The quantity of radioactive material in a package shall not exceed the relevant limits specified in paras 308–315.

Excepted packages

308. For **radioactive material** other than articles manufactured of **natural uranium, depleted uranium,** or natural thorium, an **excepted package** shall not contain activities greater than the following:

(a) Where the **radioactive material** is enclosed in or forms a component part of an instrument or other manufactured article, such as a clock or electronic apparatus, the limits specified in para. 418 for each individual item and each package, respectively; and

(b) Where the **radioactive material** is not so enclosed or manufactured, the limits specified in para. 419.

309. For articles manufactured of **natural uranium, depleted uranium,** or natural thorium, an **excepted package** may contain any quantity of such material provided that the outer surface of the uranium or thorium is enclosed in an inactive sheath made of metal or some other substantial substance.

310. For transport by post, the total activity in each **package** shall not exceed one tenth of the relevant limit specified in Table IV (Section IV).

Industrial packages

311. The total activity in a single **package** of **LSA material** or in a single **package** of **SCO** shall be so restricted that the **radiation level** specified in para. 422 shall not be exceeded, and the activity in a single **package** shall also be so restricted that the activity limits for a **conveyance** specified in para. 427 shall not be exceeded.

Type A packages

312. **Type A packages** shall not contain activities greater than the following:

(a) For **special form radioactive material** — A_1; or
(b) For all other **radioactive material** — A_2.

Values for A_1 and A_2 are listed in Tables I and II.

Type B packages

313. **Type B packages** shall not contain:

(a) Activities greater than those authorized for the **package design**,
(b) Radionuclides different from those authorized for the **package design**, or
(c) Contents in a form, or a physical or chemical state different from those authorized for the **package design**,

as specified in their certificates of approval.

Packagings containing fissile material

314. All **packagings** containing **fissile material** shall comply with the applicable activity limits for **packages** specified in paras 308–313.

315. **Packagings** containing **fissile material**, other than those containing materials which comply with the requirements of para. 560, shall not contain:

(a) A mass of **fissile material** greater than that authorized for the **package design**,
(b) Any radionuclide or **fissile material** different from those authorized for the **package design**, or
(c) Contents in a form or physical or chemical state, or in a spatial arrangement, different from those authorized for the **package design**,

as specified in their certificates of approval.

SECTION IV

PREPARATION, REQUIREMENTS AND CONTROLS FOR SHIPMENT AND FOR STORAGE IN TRANSIT

PACKAGE INSPECTION REQUIREMENTS

Before the first shipment

401. Before the first **shipment** of any **package**, the following requirements shall be fulfilled:

(a) If the design pressure of the **containment system** exceeds 35 kPa (0.35 kgf/cm^2) (gauge), it shall be ensured that the **containment system** of each **package** conforms to the approved design requirements relating to the capability of that system to maintain its integrity under pressure.
(b) For each **Type B package** and for each **packaging** containing **fissile material**, it shall be ensured that the effectiveness of its shielding and containment and, where necessary, the heat transfer characteristics, are within the limits applicable to or specified for the approved **design**.
(c) For **packagings** containing **fissile material**, where, in order to comply with the requirements of para. 559, neutron poisons are specifically included as components of the **package** tests shall be performed to confirm the presence and distribution of those neutron poisons.

Before each shipment

402. Before each **shipment** of any **package**, the following requirements shall be fulfilled:

(a) It shall be ensured that lifting attachments which do not meet the requirements of para. 506 have been removed or otherwise rendered incapable of being used for lifting the **package**.
(b) For each **Type B package** and for each **packaging** containing **fissile material**, it shall be ensured that all the requirements specified in the approval certificates and the relevant provisions of these Regulations have been satisfied.
(c) Each **Type B package** shall be held until equilibrium conditions have been approached closely enough to demonstrate compliance with the **shipment** requirements for temperature and pressure unless an exemption from these requirements has received **unilateral approval**.
(d) For each **Type B package**, it shall be ensured by examination and/or appropriate tests that all closures, valves, and other openings of the **containment**

system through which the **radioactive contents** might escape are properly closed and, where appropriate, sealed in the manner for which the demonstrations of compliance with the requirements of para. 548 were made.

TRANSPORT OF OTHER GOODS

403. A **package** shall not contain any other items except such articles and documents as are necessary for the use of the **radioactive material**. This requirement shall not preclude the transport of **low specific activity material** or **surface contaminated objects** with other items. The transport of such articles and documents in a **package**, or of **low specific activity material** or **surface contaminated objects** with other items may occur, provided that there is no interaction between them and the **packaging** or its contents that would reduce the safety of the **package**.

404. **Tanks** used for the transport of **radioactive material** shall not be used for the storage or transport of other goods.

405. The carriage of other goods with **consignments** being transported under **exclusive use** shall be permitted provided it is arranged for only by the **consignor** and it is not prohibited by other regulations.

406. **Consignments** shall be segregated from other dangerous goods during transport and storage in compliance with the relevant transport regulations for dangerous goods of each of the countries through or into which the materials will be transported, and, where applicable, with the regulations of the cognizant transport organizations, as well as these Regulations.

OTHER DANGEROUS PROPERTIES OF CONTENTS

407. In addition to the radioactive properties, any other dangerous properties of the contents of the **package**, such as explosiveness, flammability, pyrophoricity, chemical toxicity and corrosiveness, shall be taken into account in the packing, labelling, marking, placarding, storage and transport in order to be in compliance with the relevant transport regulations for dangerous goods of each of the countries through or into which the materials will be transported, and, where applicable, with the regulations of the cognizant transport organizations, as well as these Regulations.

REQUIREMENTS AND CONTROLS FOR CONTAMINATION AND FOR LEAKING PACKAGES

408. The **non-fixed contamination** on the external surfaces of a **package** shall be kept as low as practicable and, under conditions likely to be encountered in routine transport, shall not exceed the levels specified in Table III.

TABLE III. LIMITS OF NON-FIXED CONTAMINATION ON SURFACES

Type of package, overpack, freight container, tank or conveyance and its equipment	Contaminant			
	Applicable limit[a] of beta and gamma emitters and low toxicity alpha emitters		Applicable limit[a] of all other alpha emitters	
	Bq/cm^2	(μCi/cm^2)	Bq/cm^2	(μCi/cm^2)
External surfaces of: **excepted packages**	0.4	(10^{-5})	0.04	(10^{-6})
other than **excepted packages**	4	(10^{-4})	0.4	(10^{-5})
External and internal surfaces of **overpacks, freight containers** and **conveyances** and their equipment, when used in or when being prepared for the carriage of:				
— loads consisting only of **radioactive material** in packages other than **excepted packages**	4	(10^{-4})	0.4	(10^{-5})
— loads including **excepted packages** and/or non-radioactive consignments	0.4	(10^{-5})	0.04	(10^{-6})
External surfaces of **freight containers, tanks** and **conveyances** and their equipment, used in the carriage of unpackaged **radioactive material**	4	(10^{-4})	0.4	(10^{-5})

[a] The limits are applicable when averaged over any area of 300 cm^2 of any part of the surface.

409. Except as provided in para. 414, the level of **non-fixed contamination** on the external and the internal surfaces of **overpacks, freight containers** and **tanks** shall not exceed the limits specified in Table III.

410. If it is evident that a **package** is damaged or leaking, or if it is suspected that the **package** may have leaked or been damaged, access to the **package** shall be restricted and a qualified person shall, as soon as possible, assess the extent of **contamination** and the resultant **radiation level** of the **package**. The scope of the survey shall include the **package**, the **conveyance**, the adjacent loading and unloading areas, and, if necessary, all other material which has been carried in the **conveyance**. When necessary, additional steps for the protection of human health, in accordance with provisions established by the relevant **competent authority,** shall be taken to overcome and minimize the consequences of such leakage or damage.

411. **Packages** leaking **radioactive contents** in excess of allowable limits for normal conditions of transport may be removed under supervision but shall not be forwarded until repaired or reconditioned and decontaminated.

412. A **conveyance** and equipment used routinely for the carriage of **radioactive material** shall be periodically checked to determine the level of **contamination**. The frequency of such checks shall be related to the likelihood of **contamination** and the extent to which **radioactive material** is carried.

413. Except as provided in para. 414, any **conveyance**, or equipment, or part thereof which has become contaminated above the limits specified in Table III, or which shows a radiation level in excess of 5 μSv/h (0.5 mrem/h) in the course of the carriage of **radioactive material** shall be decontaminated as soon as possible by a qualified person and shall not be re-used unless the **non-fixed radioactive contamination** does not exceed the limits specified in Table III, and the **radiation level** resulting from the **fixed contamination** on surfaces after decontamination is less than 5 μSv/h (0.5 mrem/h).

414. An **overpack, freight container** or **conveyance** dedicated to the transport of **low specific activity material** or **surface contaminated objects** under **exclusive use** shall be excepted from the requirements of paras 409 and 413 solely with regard to its internal surfaces and only for as long as it remains under that specific **exclusive use.**

REQUIREMENTS AND CONTROLS FOR TRANSPORT OF EXCEPTED PACKAGES

415. **Excepted packages** shall be subject only to the following provisions in Sections IV and V:

(a) the requirements specified in paras 407, 416, 417, 436, 447 (d), 447 (l), 452 and, as applicable, 418–421;
(b) the General Requirements for all **packagings** and **packages** specified in paras 505–514;
(c) if the **excepted package** contains **fissile material**, the requirements of para. 560; and
(d) the requirements in paras 476 and 477 if transported by post.

416. The **radiation level** at any point on the external surface of an **excepted package** shall not exceed 5 μSv/h (0.5 mrem/h).

417. The **non-fixed radioactive contamination** on any external surface of an **excepted package** shall not exceed the limits specified in Table III.

418. **Radioactive material** which is enclosed in or forms a component part of an instrument or other manufactured article, with activity not exceeding the item and **package** limits specified in columns 2 and 3 respectively of Table IV, may be transported in an **excepted package** provided that:

(a) The **radiation level** at 10 cm from any point on the external surface of any unpackaged instrument or article is not greater than 0.1 mSv/h (10 mrem/h); and
(b) Each instrument or article (except radioluminescent time-pieces or devices) bears the marking "Radioactive".

TABLE IV. ACTIVITY LIMITS FOR EXCEPTED PACKAGES

Physical state of contents	Instruments and articles		Materials
	Item limits[a]	**Package** limits[a]	**Package** limits[a]
Solids:			
special form	$10^{-2} A_1$	A_1	$10^{-3} A_1$
other forms	$10^{-2} A_2$	A_2	$10^{-3} A_2$
Liquids:	$10^{-3} A_2$	$10^{-1} A_2$	$10^{-4} A_2$
Gases:			
tritium	$2 \times 10^{-2} A_2$	$2 \times 10^{-1} A_2$	$2 \times 10^{-2} A_2$
special form	$10^{-3} A_1$	$10^{-2} A_1$	$10^{-3} A_1$
other forms	$10^{-3} A_2$	$10^{-2} A_2$	$10^{-3} A_2$

[a] For mixtures of radionuclides, see paras 304–306.

419. **Radioactive material** in forms other than as specified in para. 418, with an activity not exceeding the limit specified in column 4 of Table IV, may be transported in an **excepted package** provided that:

(a) The **package** retains its contents under conditions likely to be encountered in routine transport; and

(b) The **package** bears the marking "Radioactive" on an internal surface in such a manner that a warning of the presence of **radioactive material** is visible on opening the **package**.

420. A manufactured article in which the sole **radioactive material** is unirradiated **natural uranium**, unirradiated **depleted uranium** or unirradiated natural thorium may be transported as an **excepted package** provided that the outer surface of the uranium or thorium is enclosed in an inactive sheath made of metal or some other substantial material.

Additional requirements and controls for transport of empty packagings

421. An empty **packaging** which had previously contained **radioactive material** may be transported as an **excepted package** provided that:

(a) It is in a well maintained condition and securely closed;

(b) The outer surface of any uranium or thorium in its structure is covered with an inactive sheath made of metal or some other substantial material;

(c) The level of internal **non-fixed contamination** does not exceed one thousand times the levels specified in Table III for **excepted packages**; and

(d) Any labels which may have been displayed on it in conformity with para. 440, are no longer visible.

REQUIREMENTS AND CONTROLS FOR TRANSPORT OF LSA MATERIAL AND SCO IN INDUSTRIAL PACKAGES OR UNPACKED

422. The quantity of LSA **material** or **SCO** in a single **industrial package Type 1 (IP-1), industrial package Type 2 (IP-2), industrial package Type 3 (IP-3)**, or object or collection of objects, whichever is appropriate, shall be so restricted that the external **radiation level** at 3 m from the unshielded material or object or collection of objects does not exceed 10 mSv/h (1 rem/h).

423. LSA **material** and **SCO** which is or contains **fissile material** shall meet the applicable requirements of paras 479, 480 and 559.

424. **Packages**, including **tanks** or **freight containers**, containing LSA **material** or **SCO** shall be subject to the provisions of paras 408–409.

TABLE V. INDUSTRIAL PACKAGE REQUIREMENTS FOR LSA MATERIAL AND SCO

Contents	Industrial package type	
	Exclusive use	Not under exclusive use
LSA-I[a]		
Solid	IP-1	IP-1
Liquid	IP-1	IP-2
LSA-II		
Solid	IP-2	IP-2
Liquid and gas	IP-2	IP-3
LSA-III	IP-2	IP-3
SCO-I[a]	IP-1	IP-1
SCO-II	IP-2	IP-2

[a] Under the conditions specified in para. 425, **LSA-I material** and **SCO-I** may be transported unpackaged.

425. **LSA material** and **SCO** in groups **LSA-I** and **SCO-I** may be transported unpackaged under the following conditions:

(a) All unpackaged material other than ores containing only naturally occurring radionuclides shall be transported in such a manner that under conditions likely to be encountered in routine transport there will be no escape of the contents from the **conveyance** nor will there be any loss of shielding;

(b) Each **conveyance** shall be under **exclusive use,** except when only transporting **SCO-I** on which the **contamination** on the accessible and the inaccessible surfaces is not greater than ten times the applicable level specified in para. 122; and

(c) For **SCO-I** where it is suspected that **non-fixed contamination** exists on inaccessible surfaces in excess of the values specified in para. 144(a)(i) measures shall be taken to ensure the **radioactive material** is not released into the **conveyance**.

426. **LSA material** and **SCO**, except as otherwise specified in para. 425, shall be packaged in accordance with the **package** integrity levels specified in Table V, in

TABLE VI. CONVEYANCE ACTIVITY LIMITS FOR LSA MATERIAL AND SCO IN INDUSTRIAL PACKAGES OR UNPACKAGED

Nature of material	Activity limit for **conveyances** other than by inland waterway	Activity limit for a hold or compartment of an inland water craft
LSA-I	No limit	No limit
LSA-II and **LSA-III** non-combustible solids	No limit	$100 \times A_2$
LSA-II and **LSA-III** combustible solids, and all liquids and gases	$100 \times A_2$	$10 \times A_2$
SCO	$100 \times A_2$	$10 \times A_2$

such a manner that, under conditions likely to be encountered in routine transport, there will be no escape of contents from **packages**, nor will there be any loss of shielding afforded by the **packaging**. LSA-II material, LSA-III material and SCO-II shall not be transported unpackaged.

427. The total activity in a single hold or compartment of an inland water craft, or in another conveyance, for carriage of **LSA material** or **SCO in industrial packages** or unpackaged, shall not exceed the limits shown in Table VI.

DETERMINATION OF TRANSPORT INDEX (TI)

428. The **transport index (TI)** based on radiation exposure control for a **package, overpack, tank, freight container,** or for unpackaged **LSA-I** or **SCO-I,** shall be the number derived in accordance with the following procedure:

(a) Determine the maximum **radiation level** at a distance of 1 m from the external surfaces of the **package, overpack, tank, freight container,** or unpackaged **LSA-I** and **SCO-I.** Where the **radiation level** is determined in units of millisievert per hour (mSv/h), the value determined shall be multiplied by 100. Where the **radiation level** is determined in units of millirem per hour (mrem/h), the value determined is not changed. For uranium and thorium ores and concentrates, the maximum radiation dose rate at any point 1 m from the external surface of the load may be taken as:

0.4 mSv/h (40 mrem/h)	for ores and physical concentrates of uranium and thorium
0.3 mSv/h (30 mrem/h)	for chemical concentrates of thorium
0.02 mSv/h (2 mrem/h)	for chemical concentrates of uranium, other than uranium hexafluoride.

(b) For **tanks, freight containers** and unpackaged **LSA-I** and **SCO-I**, the value determined in step (a) above shall be multiplied by the appropriate factor from Table VII.

(c) The figure obtained in steps (a) and (b) above shall be rounded up to the first decimal place (e.g. 1.13 becomes 1.2), except that a value of 0.05 or less may be considered as zero.

429. The **transport index** **(TI)** based on nuclear criticality control shall be obtained by dividing the number 50 by the value of N derived using the procedures specified in para. 567 (i.e. TI = 50/N). The value of the **transport index** for nuclear criticality control may be zero, provided that an unlimited number of **packages** is subcritical (i.e. N is effectively equal to infinity).

430. The **transport index** for each **consignment** shall be determined in accordance with Table VIII.

Additional requirements for overpacks

431. The following additional requirements shall apply to **overpacks**:

(a) **Packages** of **fissile material** for which the **transport index** for nuclear criticality control is 0 and **packages** of non-fissile **radioactive material** may be combined together in an **overpack** for transport, provided that each **package** contained therein meets the applicable requirements of these Regulations.

TABLE VII. MULTIPLICATION FACTORS FOR LARGE DIMENSION LOADS

Size of load[a]	Multiplication factor
size of load $\leq$ 1 m^2	1
1 m^2 < size of load $\leq$ 5 m^2	2
5 m^2 < size of load $\leq$ 20 m^2	3
20 m^2 < size of load	10

[a] Largest cross-sectional area of the load being measured.

TABLE VIII. DETERMINATION OF TRANSPORT INDEX

Item	Contents	Method of determining transport index (TI)
Packages	Non-**fissile material**	**TI** for radiation exposure control
	Fissile material	The larger of the **TI** for radiation exposure control or the **TI** for nuclear criticality control
Non-rigid **overpacks**	**Packages**	Sum of **TI**s of all **packages** contained
Rigid **overpacks**	**Packages**	The sum of the **TI**s of all **packages** contained, or for the original **consignor**, either the **TI** for radiation exposure control or the sum of the **TI**s of all **packages**
Freight containers	**Packages** or **overpacks**	Sum of the **TI**s of all **packages** and **overpacks** contained
	LSA material or **SCO**	Either the sum of the **TI**s or the larger of the **TI** for radiation exposure control or the **TI** for nuclear criticality control
Freight containers under **exclusive use**	**Packages** or **overpacks**	Either the sum of the **TI**s or the larger of the **TI** for radiation exposure control or the **TI** for nuclear criticality control
Tanks	**Non-fissile material**	**TI** for radiation exposure control
	Fissile material	The larger of the **TI** for radiation exposure control or the **TI** for nuclear criticality control
Unpackaged	**LSA-I** and **SCO-I**	The **TI** for radiation exposure control

(b) **Packages** of **fissile material** for which the **transport index** for nuclear criticality control exceeds 0 shall not be carried in an **overpack**.

(c) Only the original **consignor** of the **packages** contained within the **overpacks** shall be permitted to use the method of direct measurement of **radiation level** to determine the **transport index** of a rigid **overpack**.

LIMITS ON TRANSPORT INDEX AND RADIATION LEVEL FOR PACKAGES AND OVERPACKS

432. Except for **consignments** under **exclusive use**, the **transport index** of any individual **package** or **overpack** shall not exceed 10.

433. Except for **packages** or **overpacks** transported under **exclusive use** by rail and by road under the conditions specified in para. 469(a), or under **exclusive use** and **special arrangement** by **vessel** or by air under the conditions specified in paras 471 or 475 respectively, the maximum **radiation level** at any point on any external surface of a **package** or **overpack** shall not exceed 2 mSv/h (200 mrem/h).

434. The maximum **radiation level** at any point on any external surface of a **package** under **exclusive use** shall not exceed 10 mSv/h (1000 mrem/h).

CATEGORIES

435. **Packages** and **overpacks** shall be assigned to either category I-WHITE, II-YELLOW or III-YELLOW in accordance with the conditions specified in Tables IX and X, as applicable, and with the following requirements:

(a) For a **package**, both the **transport index** and the surface **radiation level** conditions shall be taken into account in determining which is the appropriate category. Where the **transport index** satisfies the condition for one category but the surface **radiation level** satisfies the condition for a different category, the **package** shall be assigned to the higher category of the two. For this purpose, category I-WHITE shall be regarded as the lowest category.

(b) The **transport index** shall be determined following the procedures specified in paras 428–430, and subject to the limitation of para. 431(c).

(c) If the **transport index** is greater than 10, the **package** or **overpack** shall be transported under **exclusive use**.

(d) If the surface **radiation level** is greater than 2 mSv/h (200 mrem/h), the **package** or **overpack** shall be transported under **exclusive use** and under the provisions of paras 469(a), 471 and 475, as appropriate.

TABLE IX. CATEGORIES OF PACKAGES

Transport index	Conditions — Maximum **radiation level** at any point on external surface	Category
0^a	Not more than 0.005 mSv/h (0.5 mrem/h)	I-WHITE
More than 0 but not more than 1^a	More than 0.005 mSv/h (0.5 mrem/h) but not more than 0.5 mSv/h (50 mrem/h)	II-YELLOW
More than 1 but not more than 10	More than 0.5 mSv/h (50 mrem/h) but not more than 2 mSv/h (200 mrem/h)	III-YELLOW
More than 10	More than 2 mSv/h (200 mrem/h) but not more than 10 mSv/h (1000 mrem/h)	III-YELLOW and also under **exclusive use**

[a] If the measured **TI** is not greater than 0.05, the value quoted may be zero in accordance with para. 428(c).

TABLE X. CATEGORIES OF OVERPACKS INCLUDING FREIGHT CONTAINERS WHEN USED AS OVERPACKS

Transport index	Category
0	I-WHITE
TI greater than 0 but less than or equal to 1	II-YELLOW
TI greater than 1	III-YELLOW

(e) A **package** transported under a **special arrangement** shall be assigned to category III-YELLOW.
(f) An **overpack** which contains **packages** transported under **special arrangements** shall be assigned to category III-YELLOW.

MARKING, LABELLING AND PLACARDING

Marking

436. Each **package** of gross mass exceeding 50 kg shall have its permissible gross mass legibly and durably marked on the outside of the **packaging**.

437. Each **package** which conforms to a **Type A package design** shall be legibly and durably marked on the outside of the **packaging** with "Type A".

438. Each **package** which conforms to a **design** approved under paras 704–714 shall be legibly and durably marked on the outside of the **packaging** with:

(a) The identification mark allocated to that **design** by the **competent authority**;
(b) A serial number to uniquely identify each **packaging** which conforms to that **design**; and
(c) In the case of a **Type B(U) or Type B(M) package design**, with "Type B(U)" or "Type B(M)".

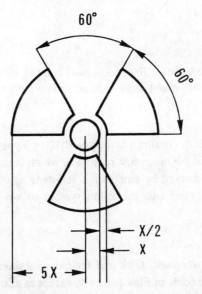

FIG. 1. Basic trefoil symbol with proportions based on a central circle of radius X. The minimum size of X shall be 4 mm.

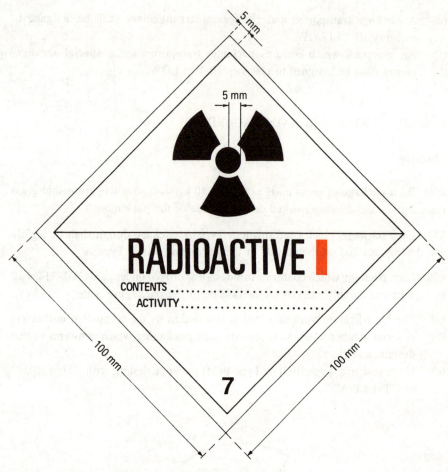

FIG. 2. Category I-WHITE label. The background colour of the label shall be white, the colour of the trefoil and the printing shall be black, and the colour of the category bar shall be red.

439. Each **package** which conforms to a **Type B(U)** or **Type B(M) package design** shall have the outside of the outermost receptacle which is resistant to the effects of fire and water plainly marked by embossing, stamping or other means resistant to the effects of fire and water with the trefoil symbol shown in Fig. 1.

Labelling

440. Each **package**, **overpack**, **tank** and **freight container** shall bear the labels which conform to the models in Figs 2, 3 or 4, except as allowed under the alternative provision of para. 443 for large **freight containers** and **tanks**, according to the appropriate category. Any labels which do not relate to the contents shall be removed

FIG. 3. Category II-YELLOW label. The background colour of the upper half of the label shall be yellow and of the lower half white, the colour of the trefoil and the printing shall be black, and the colour of the category bars shall be red.

or covered. For **radioactive material** having other dangerous properties see para. 407.

441. The labels shall be affixed to two opposite sides of the outside of a **package** or **overpack**, or on the outside of all four sides of a **freight container** or **tank**.

442. Each label shall be completed with the following information:

(a) Contents:
    (i) Except for **LSA-I material**, the name of the radionuclide as taken from Table I, using the symbols prescribed therein. For mixtures of radionuclides, the most restrictive nuclides must be listed to the extent the

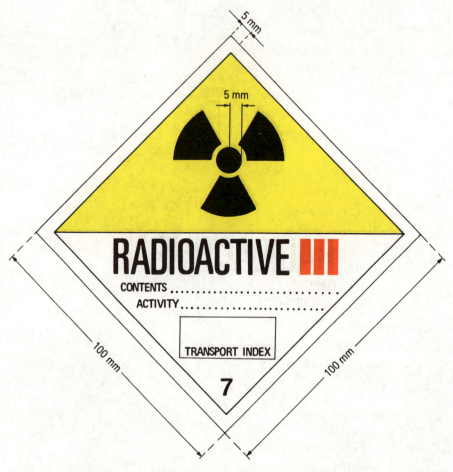

FIG. 4. Category III-YELLOW label. The background colour of the upper half of the label shall be yellow and of the lower half white, the colour of the trefoil and the printing shall be black, and the colour of the category bars shall be red.

space on the line permits. The group of **LSA** or **SCO** shall be shown following the name of the radionuclide. The terms "LSA-II", "LSA-III", "SCO-I" and "SCO-II" shall be used for this purpose.

(ii) For **LSA-I** materials, the term "LSA-I" is all that is necessary; the name of the radionuclide is not necessary.

(b) Activity: The maximum activity of the **radioactive contents** during transport expressed in units of becquerels (Bq) (or curies (Ci)) with the appropriate SI prefix (see Appendix II). For **fissile material**, the total mass in units of grams (g), or multiples thereof, may be used in place of activity.

FIG. 5. Placard. Minimum dimensions are given; when larger dimensions are used the relative proportions must be maintained. The figure '7' shall not be less than 25 mm high. The background colour of the upper half of the placard shall be yellow and of the lower half white, the colour of the trefoil and the printing shall be black. The use of the word "RADIOACTIVE" in the bottom half is optional to allow the alternative use of this placard to display the appropriate United Nations Number for the **consignment**.

(c) For **overpacks**, **tanks** and **freight containers**, the "contents" and "activity" entries on the label shall bear the information required in paras 442(a) and 442(b), respectively, totalled together for the entire contents of the **overpack**, **tank**, or **freight container** except that on labels for **overpacks** or **freight containers** containing mixed loads of **packages** with different radionuclides, such entries may read "See Transport Documents".

(d) **Transport index**: See para. 430. (No **transport index** entry required for category I-WHITE.)

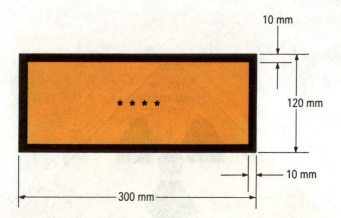

FIG. 6. *Placard for separate display of the United Nations Number. The background colour of the placard shall be orange and the border and United Nations Number shall be black. The symbol "****" denotes the space in which the appropriate United Nations Number for **radioactive material**, as specified in Appendix I, shall be displayed.*

Placarding

443. Large **freight containers** carrying **packages** other than **excepted packages**, and **tanks** shall bear four placards which conform with the model given in Fig. 5. The placards shall be affixed in a vertical orientation to each side wall and each end wall of the **freight container** or **tank**. Any placards which do not relate to the contents shall be removed. Instead of using a label and a placard, it is permitted as an alternative to use enlarged labels only, as shown in Figs 2, 3 and 4, with dimensions of the minimum size shown in Fig. 5.

444. Where the **consignment** in the **freight container** or **tank** is unpackaged **LSA-I** or **SCO-I** or where an **exclusive use consignment** in a **freight container** is packaged **radioactive material** comprised of a single United Nations Number commodity, the appropriate United Nations Number for the **consignment** (see Appendix I) shall also be displayed, in black digits not less than 65 mm high, either:

(a) in the lower half of the placard shown in Fig. 5, against the white background, or
(b) on the placard shown in Fig. 6.

When the alternative given in (b) above is used, the subsidiary placard shall be affixed immediately adjacent to the main placard, on all four sides of the **freight container** or **tank**.

Design of labels and placards

445. The labels and placards required by these Regulations shall conform to the appropriate models shown in Figs 1–6 and shall conform to the colours shown in Figs 2–6.

CONSIGNOR'S RESPONSIBILITIES

446. Compliance with the requirements of paras 421(d) and 436–444 for labelling, marking and placarding shall be the responsibility of the **consignor**.

Particulars of consignment

447. The **consignor** shall include in the transport documents with each **consignment** the following information, as applicable, in the order given:

(a) The proper shipping name, as specified in Appendix I;
(b) The United Nations Class Number "7";
(c) The words "RADIOACTIVE MATERIAL" unless these words are contained in the proper shipping name;
(d) The United Nations Number assigned to the material as specified in Appendix I;
(e) For **LSA material**, the group notation "LSA-I", "LSA-II" or "LSA-III", as appropriate;
(f) For **SCO**, the group notation "SCO-I" or "SCO-II", as appropriate;
(g) The name or symbol of each radionuclide or, for mixtures of radionuclides, an appropriate general description or a list of the most restrictive nuclides;
(h) A description of the physical and chemical form of the material, or a notation that the material is **special form radioactive material**. A generic chemical description is acceptable for chemical form;
(i) The maximum activity of the **radioactive contents** during transport expressed in units of becquerels (Bq) (or curies (Ci)) with an appropriate SI prefix (see Appendix II). For **fissile material**, the total mass of **fissile material** in units of grams (g), or appropriate multiples thereof, may be used in place of activity;
(j) The category of the **package**, i.e. I-WHITE, II-YELLOW, III-YELLOW;
(k) The **transport index** (categories II-YELLOW and III-YELLOW only);
(l) All items and materials transported under the provisions for **excepted packages** (see paras 415–421) shall be described in the transport document as "RADIOACTIVE MATERIAL, EXCEPTED PACKAGE", and shall include the proper shipping name of the substance or article being transported from the list of United Nations Numbers (see Appendix I);

(m) For a **consignment** of **fissile material**, where all of the **packages** in the **consignment** are excepted under para. 560, the words "FISSILE EXCEPTED";

(n) The identification mark for each **competent authority** approval certificate (**special form radioactive material, special arrangement, package design,** or **shipment**) applicable to the **consignment**;

(o) For **consignments** of **packages** in an **overpack** or **freight container**, a detailed statement of the contents of each **package** within the **overpack** or **freight container** and, where appropriate, of each **overpack** or **freight container** in the **consignment**. If **packages** are to be removed from the **overpack** or **freight container** at a point of intermediate unloading, appropriate transport documentation shall be made available;

(p) Where a **consignment** is required to be shipped under **exclusive use**, the statement "EXCLUSIVE USE SHIPMENT".

Consignor's declaration

448. The **consignor** shall include in the transport documents a declaration in the following terms or in terms having an equivalent meaning:

"I hereby declare that the contents of this consignment are fully and accurately described above by proper shipping name and are classified, packed, marked and labelled, and are in all respects in proper condition for transport by (insert mode(s) of transport involved) according to the applicable international and national governmental regulations."

449. If the intent of the declaration is already a condition of transport within a particular international convention, the **consignor** need not produce such a declaration for that part of the transport covered by the convention.

450. The declaration shall be signed and dated by the **consignor**. Facsimile signatures are authorized where applicable laws and regulations recognize the legal validity of facsimile signatures.

451. The declaration shall be made on the same document which contains the particulars of **consignment** listed in para. 447.

Removal or covering of labels

452. When an empty **packaging** is shipped as an **excepted package** under the provisions of para. 421, the previously displayed labels shall not be visible.

Information for carriers

453. The **consignor** shall provide in the transport documents a statement regarding actions, if any, that are required to be taken by the **carrier**. The statement shall be in the languages deemed necessary by the **carrier** or the authorities concerned, and shall include at least the following points:

(a) Supplementary operational requirements for loading, stowage, transport, handling and unloading of the **package, overpack, freight container** or **tank** including any special stowage provisions for the safe dissipation of heat (see para. 463), or a statement that no such requirements are necessary;
(b) Restrictions on the mode of transport or **conveyance** and any necessary routing instructions;
(c) Emergency arrangements appropriate to the **consignment**.

454. The applicable **competent authority** certificates need not necessarily accompany the **consignment**. The **consignor** shall, however, be prepared to provide them to the **carrier** before loading, unloading and any transshipment.

Notification of competent authorities

455. Before the first **shipment** of any **package** requiring **competent authority** approval, the **consignor** shall ensure that copies of each applicable **competent authority** certificate applying to that **package design** have been submitted to the **competent authority** of each country through or into which the **consignment** is to be transported. The **consignor** is not required to await an acknowledgement from the **competent authority**, nor is the **competent authority** required to make such acknowledgement of receipt of the certificate.

456. For each **shipment** listed in (a), (b) or (c) below, the **consignor** shall notify the **competent authority** of each country through or into which the **consignment** is to be transported. This notification shall be in the hands of each **competent authority** prior to the commencement of the **shipment**, and preferably at least 7 days in advance.

(a) **Type B(U) packages** containing **radioactive material** with an activity greater than 3×10^3 A_1 or 3×10^3 A_2, as appropriate, or 1000 TBq (20 kCi), whichever is the lower;
(b) **Type B(M) packages**;
(c) Transport under **special arrangement**.

457. The **consignment** notification shall include:

(a) Sufficient information to enable the identification of the **package** including all applicable certificate numbers and identification marks;

(b) Information on the date of **shipment**, the expected date of arrival and proposed routing;
(c) The name of the **radioactive material** or nuclide;
(d) A description of the physical and chemical form of the **radioactive material**, or whether it is **special form radioactive material**;
(e) The maximum activity of the **radioactive contents** during transport expressed in units of becquerels (Bq) (or curies (Ci)) with an appropriate SI prefix (see Appendix II). For **fissile material**, the mass of **fissile material** in units of grams (g), or multiples thereof, may be used in place of activity.

458. The **consignor** is not required to send a separate notification if the required information has been included in the application for **shipment** approval. See para. 718.

Possession of certificates and operating instructions

459. The **consignor** shall have in his or her possession a copy of each certificate required under Section VII and a copy of the instructions with regard to the proper closing of the **package** and other preparations for **shipment** before making any **shipment** under the terms of the certificates.

TRANSPORT

Segregation during transport

460. **Packages, overpacks, freight containers** and **tanks** shall be segregated during transport:

(a) from places occupied by workers and members of the public and from undeveloped photographic film, for radiation exposure control purposes, in accordance with paras 205 and 206, and
(b) from other dangerous goods in accordance with para. 406.

461. Categories II-YELLOW or III-YELLOW **packages** or **overpacks** shall not be carried in compartments occupied by passengers, except those exclusively reserved for couriers specially authorized to accompany such **packages** or **overpacks**.

Stowing for transport

462. **Consignments** shall be securely stowed.

463. Provided that its average surface heat flux does not exceed 15 W/m^2 and that the immediately surrounding cargo is not in sacks or bags, a **package** or **overpack** may be carried among packaged general cargo without any special stowage provi-

sions except as may be specifically required by the **competent authority** in an applicable approval certificate.

464. Except in the case of **shipment** under **special arrangement, mixing of packages** of different kinds of **radioactive material**, including **fissile material**, and mixing of different kinds of **packages** with different **transport indexes** is permitted without specific **competent authority** approval. In the case of **shipments** under **special arrangement**, mixing shall not be permitted except as specifically authorized under the **special arrangement**.

465. Loading of **tanks** and **freight containers** and accumulation of **packages, overpacks, tanks** and **freight containers** shall be controlled as follows:

(a) The total number of **packages, overpacks, tanks** and **freight containers** aboard a single **conveyance** shall be so limited that the total sum of the **transport indexes** aboard the **conveyance** does not exceed the values shown in Table XI. For **consignments** of **LSA-I material** there shall be no limit on the sum of the **transport indexes**.

(b) The **radiation level** under conditions likely to be encountered in routine transport shall not exceed 2 mSv/h (200 mrem/h) at any point on, and 0.1 mSv/h (10 mrem/h) at 2 m from, the external surface of the **conveyance**.

466. Any **package** or **overpack** having a **transport index** greater than 10 shall be transported only under **exclusive use**.

Additional requirements relating to transport by rail and by road

467. Rail and road **vehicles** carrying **packages, overpacks, tanks** or **freight containers** labelled with any of the labels shown in Figs 2, 3 or 4, or carrying **consignments** under **exclusive use**, shall display the placard shown in Fig. 5 on each of:

(a) The two external lateral walls in the case of a rail **vehicle**;
(b) The two external lateral walls and the external rear wall in the case of a road **vehicle**.

In the case of a **vehicle** without sides the placards may be affixed directly on the cargo-carrying unit provided that they are readily visible; in the case of physically large **tanks** or **freight containers**, the placards on the **tanks** or **freight containers** shall suffice. Any placards which do not relate to the contents shall be removed.

468. Where the **consignment** in or on the **vehicle** is unpackaged **LSA-I** or **SCO-I** or where an **exclusive use consignment** is packaged **radioactive material** comprised of a single United Nations Number commodity, the appropriate United Nations Number (see Appendix I) shall also be displayed, in black digits not less than 65 mm high, either:

(a) In the lower half of the placard shown in Fig. 5, against the white background, or
(b) On the placard shown in Fig. 6.

TABLE XI. TI LIMITS FOR FREIGHT CONTAINERS AND CONVEYANCES

Type of **freight container** or **conveyance**	Limit on total sum of **transport indexes** in a single **freight container** or aboard a **conveyance**			
	Not under **exclusive use**		Under **exclusive use**	
	Non-fissile material	Fissile material	Non-fissile material	Fissile material
Freight container — Small	50	50	n.a.	n.a.
Freight container — Large	50	50	No limit	100
Vehicle	50	50	No limit	100
Aircraft				
passenger	50	50	n.a.	n.a.
cargo	200	50	No limit	100
Inland waterway **vessel**	50	50	No limit	100
Seagoing **vessel**[a]				
1. Hold, compartment or **defined deck area:**				
packages, overpacks, small **freight containers**	50	50	No limit	100
large **freight containers**	200[b]	50	No limit	100
2. Total **vessel**:				
packages, etc.	200[b]	200[b]	No limit	200[c]
large **freight containers**	No limit[b]	No limit[b]	No limit	No limit[c]

[a] **Packages** or **overpacks** carried in or on a **vehicle** which are in accordance with the provisions of para. 469 may be transported by vessels provided that they are not removed from the vehicle at any time while on board the vessel.

[b] The **consignment** shall be so handled and stowed that the sum of **TIs** in any individual group does not exceed 50, and that each group is handled and stowed such that the groups are separated from each other by at least 6 m.

[c] The **consignment** shall be so handled and stowed that the sum of **TIs** in any individual group does not exceed 100, and that each group is handled and stowed such that the groups are separated from each other by at least 6 m. The intervening space between groups may be occupied by other cargo in accordance with para. 405.

When the alternative given in (b) above is used, the subsidiary placard shall be affixed immediately adjacent to the main placard, on either the two lateral walls in the case of a rail **vehicle** or the two lateral walls and the end wall in the case of a road **vehicle**.

469. For **consignments** under **exclusive use**, the **radiation level** shall not exceed:

(a) 10 mSv/h (1000 mrem/h) at any point on the external surface of any **package** or **overpack**, and may only exceed 2 mSv/h (200 mrem/h) provided that:
  (i) the **vehicle** is equipped with an enclosure which, during routine transport, prevents the access of unauthorized persons to the interior of the enclosure and
  (ii) provisions are made to secure the **package** or **overpack** so that its position within the **vehicle** remains fixed during routine transport and
  (iii) there are no loading or unloading operations between the beginning and end of the **shipment**;

(b) 2 mSv/h (200 mrem/h) at any point on the outer surfaces of the **vehicle**, including the upper and lower surfaces, or in the case of an open **vehicle**, at any point on the vertical planes projected from the outer edges of the **vehicle**, on the upper surface of the load, and on the lower external surface of the **vehicle**; and

(c) 0.1 mSv/h (10 mrem/h) at any point 2 m from the vertical planes represented by the outer lateral surfaces of the **vehicle**, or, if the load is transported in an open **vehicle**, at any point 2 m from the vertical planes projected from the outer edges of the **vehicle**.

470. In the case of road **vehicles**,

(a) No persons other than the driver and assistants shall be permitted in **vehicles** carrying **packages**, **overpacks**, **tanks** or **freight containers** bearing category II-YELLOW or III-YELLOW labels; and

(b) The **radiation level** at any normally occupied position shall not exceed 0.02 mSv/h (2 mrem/h) unless the persons occupying such positions are provided with personal monitoring devices.

Additional requirements relating to transport by vessels

471. **Packages** having a surface **radiation level** greater than 2 mSv/h (200 mrem/h), unless being carried in or on a **vehicle** under **exclusive use** in accordance with Table XI, footnote (a), shall not be transported by **vessel** except under **special arrangement**.

472. The transport of **consignments** by means of a special use **vessel** which, by virtue of its design, or by reason of its being chartered, is dedicated to the purpose of carrying **radioactive material**, shall be excepted from the requirements specified in para. 465 provided that the following conditions are met:

(a) A radiation protection programme for the **shipment** shall be prepared and shall be approved by the **competent authority** of the flag state of the **vessel** and, when requested, by the **competent authority** at each port of call;
(b) Stowage arrangements shall be predetermined for the whole voyage, including any **consignments** to be loaded at ports of call en route; and
(c) The loading, handling and stowage and the unloading of the **consignments** shall be supervised by persons qualified in the carriage of **radioactive material**.

Additional requirements relating to transport by air

473. **Type B(M) packages** and **consignments** under **exclusive use** shall not be transported on **passenger aircraft**.

474. Vented **Type B(M) packages, packages** which require external cooling by an ancillary cooling system, **packages** subject to operational controls during transport, and **packages** containing liquid pyrophoric materials shall not be transported by air.

475. **Packages** having a surface **radiation level** greater than 2 mSv/h (200 mrem/h), otherwise allowed under **exclusive use** transport by road or rail, shall not be transported by air except by **special arrangement**.

Additional requirements relating to transport by post

476. A **consignment** that conforms with the requirements of para. 415, and in which the activity of the contents does not exceed one tenth of the limits prescribed in Table IV, may be accepted for domestic movement by national postal authorities, subject to such additional requirements as those authorities may prescribe.

477. A **consignment** that conforms with the requirements of para. 415, and in which the activity of the contents does not exceed one tenth of the limits prescribed in Table IV, may be accepted for international movement by post, subject in particular to the following additional requirements as prescribed by the Acts of the Universal Postal Union:

(a) It shall be deposited with the postal service only by **consignors** authorized by the national authority;
(b) It shall be dispatched by the quickest route, normally by air;
(c) It shall be plainly and durably marked on the outside with the words "RADIOACTIVE MATERIAL — Quantities permitted for Movement by Post"; these words shall be crossed out if the **packaging** is returned empty;
(d) It shall carry on the outside the name and address of the **consignor** with the request that the **consignment** be returned in the case of non-delivery; and
(e) The name and address of the **consignor** and the contents of the **consignment** shall be indicated on the internal **packaging**.

STORAGE IN TRANSIT

478. **Packages, overpacks, freight containers** and **tanks** shall be segregated during storage in transit:

(a) From places occupied by workers and members of the public and from undeveloped photographic film, for radiation exposure control purposes, in accordance with paras 205 and 206; and
(b) From other dangerous goods in accordance with para. 406.

479. The number of category II-YELLOW and category III-YELLOW **packages, overpacks, tanks** and **freight containers** stored in any one storage area, such as a transit area, terminal building, store-room or assembly yard, shall be so limited that the total sum of the **transport indexes** in any individual group of such **packages, overpacks, tanks** or **freight containers** does not exceed 50. Groups of such **packages, overpacks, tanks** and **freight containers** shall be stored so as to maintain a spacing of at least 6 m from other groups of such **packages, overpacks, tanks** or **freight containers**.

480. Where the **transport index** of a single **package, overpack, tank** or **freight container** exceeds 50 or the total **transport index** on board a **conveyance** exceeds 50, as permitted in Table XI, storage shall be such as to maintain a spacing of at least 6 m from other groups of **packages, overpacks, tanks** or **freight containers** or other **conveyance** carrying **radioactive material**.

481. **Consignments** in which the **radioactive contents** are **LSA-I materials** shall be excepted from the requirements of paras 479 and 480.

482. Except in the case of **shipment** under **special arrangement**, mixing of **packages** of different kinds of **radioactive material**, including **fissile material**, and mixing of different kinds of **packages** with different **transport indexes** is permitted without specific **competent authority** approval. In the case of **shipment** under **special arrangement**, mixing shall not be permitted except as specifically authorized under the **special arrangement**.

CUSTOMS OPERATIONS

483. Customs operations involving examination of the **radioactive contents** of a **package** shall be carried out only in a place where adequate means of controlling radiation exposure are provided and in the presence of qualified persons. Any **package** opened on customs instructions shall, before being forwarded to the **consignee**, be restored to its original condition.

UNDELIVERABLE CONSIGNMENTS

484. Where a **consignment** is undeliverable the **consignment** shall be placed in a safe location and the appropriate **competent authority** shall be informed as soon as possible and a request made for instructions on further action.

Section V

REQUIREMENTS FOR RADIOACTIVE MATERIALS AND FOR PACKAGINGS AND PACKAGES

REQUIREMENTS FOR RADIOACTIVE MATERIALS

Requirements for LSA-III material

501. **LSA-III material** shall be a solid of such a nature that if the entire contents of a **package** were subjected to the test specified in para. 603 the activity in the water would not exceed 0.1 A_2.

Requirements for special form radioactive material

502. **Special form radioactive material** shall have at least one dimension not less than 5 mm.

503. **Special form radioactive material** shall be of such a nature or shall be so designed that if it is subjected to the tests specified in paras 604–613, it shall meet the following requirements:

(a) It would not break or shatter under the impact, percussion and bending tests in paras 607, 608, 609 and 611(a) as applicable;
(b) It would not melt or disperse in the heat test in paras 610 and 611(b) as applicable; and
(c) The activity in the water from the leaching tests specified in paras 612 and 613 would not exceed 2 kBq (50 nCi); or alternatively for sealed sources, the leakage rate for the volumetric leakage assessment test specified in the International Organization for Standardization Document ISO/TR 4826-1979(E), Sealed Radioactive Sources — Leak Test Methods, would not exceed the applicable acceptance threshold acceptable to the **competent authority**.

504. When a sealed capsule constitutes part of the **special form radioactive material**, the capsule shall be so constructed that it can be opened only by destroying it.

GENERAL REQUIREMENTS FOR ALL PACKAGINGS AND PACKAGES

505. The **package** shall be so designed in relation to its mass, volume and shape that it can be easily and safely handled and transported. In addition, the **package** shall be so designed that it can be properly secured in or on the **conveyance** during transport.

506. The **design** shall be such that any lifting attachments on the **package** will not fail when used in the intended manner and that, if failure of the attachments should occur, the ability of the **package** to meet other requirements of these Regulations would not be impaired. Assessment shall take account of appropriate safety factors to cover snatch lifting.

507. Attachments and any other features on the outer surface of the **package** which could be used to lift it shall be designed either to support its mass in accordance with the requirements of para. 506 or shall be removable or otherwise rendered incapable of being used during transport.

508. As far as practicable, the **packaging** shall be so designed and finished that the external surfaces are free from protruding features and can be easily decontaminated.

509. As far as practicable, the outer layer of the **package** shall be so designed as to prevent the collection and the retention of water.

510. Any features added to the **package** at the time of transport which are not part of the **package** shall not reduce its safety.

511. The **package** shall be capable of withstanding the effects of any acceleration, vibration or vibration resonance which may arise under conditions likely to be encountered in routine transport without any deterioration in the effectiveness of the closing devices on the various receptacles or in the integrity of the **package** as a whole. In particular, nuts, bolts and other securing devices shall be so designed as to prevent them from becoming loose or being released unintentionally, even after repeated use.

512. The materials of the **packaging** and any components or structures shall be physically and chemically compatible with each other and with the **radioactive contents**. Account shall be taken of their behaviour under irradiation.

513. All valves through which the **radioactive contents** could otherwise escape shall be protected against unauthorized operation.

514. For **radioactive material** having other dangerous properties, the **package design** shall take into account those properties (see paras 105 and 407).

ADDITIONAL REQUIREMENTS FOR PACKAGES TRANSPORTED BY AIR

515. For **packages** to be transported by air, the temperature of the accessible surfaces shall not exceed 50°C at an ambient temperature of 38°C with no account taken for insolation.

516. **Packages** to be transported by air shall be so designed that, if they were exposed to ambient temperatures ranging from $-40°C$ to $+55°C$, the integrity of containment would not be impaired.

517. **Packages** containing liquid **radioactive materials** to be transported by air, shall be capable of withstanding without leakage an internal pressure which produces a pressure differential of not less than 95 kPa (0.95 kgf/cm^2).

REQUIREMENTS FOR INDUSTRIAL PACKAGES

Requirements for industrial package Type 1 (IP-1)

518. An **industrial package Type 1 (IP-1)** shall be designed to meet the requirements specified in paras 505–514 and 525, and, in addition, the requirements of paras 515–517 if carried by air.

Additional requirements for industrial package Type 2 (IP-2)

519. A **package**, to be qualified as an **industrial package Type 2 (IP-2)**, shall be designed to meet the requirements for **IP-1** as specified in para. 518 and, in addition, if it were subjected to the tests specified in paras 622 and 623, or, alternatively to the tests specified for packaging group III in the "Recommendations on the Transport of Dangerous Goods", prepared by the United Nations Committee of Experts on the Transport of Dangerous Goods, it would prevent:

(a) The loss or dispersal of the **radioactive contents**; and
(b) The loss of shielding integrity which would result in more than a 20% increase in the **radiation level** at any external surface of the **package**.

Additional requirements for industrial package Type 3 (IP-3)

520. A **package**, to be qualified as an **industrial package Type 3 (IP-3)**, shall be designed to meet the requirements for **IP-1** as specified in para. 518 and, in addition, the requirements specified in paras 525–538.

Alternative requirements for tanks and freight containers to qualify as IP-2 and IP-3

521. Tank containers may also be used as **industrial package Types 2** and **3**, **(IP-2)** and **(IP-3)**, provided that:

(a) They satisfy the requirements for **IP-1** specified in para. 518;
(b) They shall be designed to conform to the standards prescribed in Chapter 12 of the "Recommendations on the Transport of Dangerous Goods" prepared by the United Nations Committee of Experts on the Transport of Dangerous Goods, or other requirements at least equivalent to those standards, and are capable of withstanding a test pressure of 265 kPa (2.65 kgf/cm^2); and

(c)   They shall be designed so that any additional shielding which is provided shall be capable of withstanding the static and dynamic stresses resulting from normal handling and routine conditions of transport and of preventing a loss of shielding which would result in more than a 20% increase in the **radiation level** at any external surface of the tank containers.

522. **Tanks,** other than tank containers, may also be used as **industrial package Types 2** and **3, (IP-2)** and **(IP-3)**, for transporting **LSA-I** and **LSA-II** liquids and gases as prescribed in Table V, provided that they conform to standards at least equivalent to those prescribed in para. 521.

523. **Freight containers** may also be used as **industrial package Types 2** or **3, (IP-2)** and **(IP-3)**, provided that:

(a)   They shall satisfy the requirements for **IP-1** specified in para. 518; and
(b)   They shall be designed to conform to the requirements prescribed in the International Organization for Standardization document ISO 1496/1-1978, "Series 1 Freight Containers — Specifications and Testing — Part 1: General Cargo Containers", and if they were subjected to the tests prescribed in that document they would prevent:
   (i)   loss or dispersal of the **radioactive contents**; and
   (ii) loss of shielding which would result in more than a 20% increase in the **radiation level** at any external surface of the **freight containers**.

REQUIREMENTS FOR TYPE A PACKAGES

524. **Type A packages** shall be designed to meet the requirements specified in paras 505–514 and, in addition, the requirements of paras 515–517 if carried by air, and of paras 525–540.

525. The smallest overall external dimension of the **package** shall not be less than 10 cm.

526. The outside of the **package** shall incorporate a feature such as a seal, which is not readily breakable and which, while intact, will be evidence that it has not been opened.

527. Any tie-down attachments on the **package** shall be so designed that, under both normal and accident conditions, the forces in those attachments shall not impair the ability of the **package** to meet the requirements of the Regulations.

528. The design of the **package** shall take into account temperatures ranging from –40°C to +70°C for the components of the **packaging**. Special attention shall be given to freezing temperatures for liquid contents and to the potential degradation of **packaging** materials within the given temperature range.

529. The **design**, fabrication and manufacturing techniques shall be in accordance with national or international standards, or other requirements, acceptable to the **competent authority**.

530. The **design** shall include a **containment system** securely closed by a positive fastening device which cannot be opened unintentionally or by a pressure which may arise within the **package**.

531. **Special form radioactive material** may be considered as a component of the **containment system**.

532. If the **containment system** forms a separate unit of the **package**, it shall be capable of being securely closed by a positive fastening device which is independent of any other part of the **packaging**.

533. The **design** of any component of the **containment system** shall take into account, where applicable, the radiolytic decomposition of liquids and other vulnerable materials and the generation of gas by chemical reaction and radiolysis.

534. The **containment system** shall retain its **radioactive contents** under a reduction of ambient pressure to 25 kPa (0.25 kgf/cm^2).

535. All valves, other than pressure relief valves, shall be provided with an enclosure to retain any leakage from the valve.

536. A radiation shield which encloses a component of the **package** specified as a part of the **containment system** shall be so designed as to prevent the unintentional release of that component from the shield. Where the radiation shield and such component within it form a separate unit, the radiation shield shall be capable of being securely closed by a positive fastening device which is independent of any other **packaging** structure.

537. A **package** shall be so designed that if it were subjected to the tests specified in paras 619–624, it would prevent:

(a) Loss or dispersal of the **radioactive contents**; and
(b) Loss of shielding integrity which would result in more than a 20% increase in the **radiation level** at any external surface of the **package**.

538. The **design** of a **package** intended for liquid **radioactive material** shall make provision for ullage to accommodate variations in the temperature of the contents, dynamic effects and filling dynamics.

539. A **Type A package** designed to contain liquids shall, in addition:

(a) Be adequate to meet the conditions specified in para. 537 if the **package** is subjected to the tests specified in para. 625; and

(b) Either
   (i) Be provided with sufficient absorbent material to absorb twice the volume of the liquid contents. Such absorbent material must be suitably positioned so as to contact the liquid in the event of leakage; or
   (ii) Be provided with a **containment system** composed of primary inner and secondary outer containment components designed to ensure retention of the liquid contents, within the secondary outer containment components, even if the primary inner components leak.

However, the requirements given in paras 539(b) shall not apply in the case of a **Type B package** designed and approved for liquids which contains the same liquids having an activity equal to or less than the A_2 limit for the authorized contents.

540. A **package** designed for compressed gases or **uncompressed gases** shall prevent loss or dispersal of the **radioactive contents** if the **package** were subjected to the tests specified in para. 625. A **package** designed for contents not exceeding 40 TBq (1000 Ci) of tritium or for noble gases in gaseous form with contents not exceeding A_2 shall be excepted from this requirement.

REQUIREMENTS FOR TYPE B PACKAGES

541. **Type B packages** shall be designed to meet the requirements specified in paras 505–514, the requirements of paras 515–517 if carried by air, and of paras 525–538, except as specified in para. 548(a), and, in addition, the requirements specified in paras 542–548 and paras 550–556 or paras 557 and 558, as applicable.

542. A **package** shall be so designed that, if it were subjected to the tests in paras 626–629, it would retain sufficient shielding to ensure that the **radiation level** at 1 m from the surface of the **package** would not exceed 10 mSv/h (1 rem/h) with the maximum **radioactive contents** which the **package** is designed to carry.

543. A **package** shall be so designed that, under the ambient conditions specified in paras 545 and 546, heat generated within the **package** by the **radioactive contents** shall not, under normal conditions of transport, as demonstrated by the tests in paras 619–624, adversely affect the **package** in such a way that it would fail to meet the applicable requirements for containment and shielding if left unattended for a period of one week. Particular attention shall be paid to the effects of heat which may:

(a) Alter the arrangement, the geometrical form or the physical state of the **radioactive contents** or, if the **radioactive material** is enclosed in a can or receptacle (for example, clad fuel elements), cause the can, receptacle or **radioactive material** to deform or melt; or

TABLE XII. INSOLATION DATA

Form and location of surface	Insolation for 12 hours per day (W/m^2)
Flat surfaces transported horizontally:	
— base	none
— other surfaces	800
Flat surfaces not transported horizontally:	
— each surface	200[a]
Curved surfaces	400[a]

[a] Alternatively, a sine function may be used, with an absorption coefficient adopted and the effects of possible reflection from neighbouring objects neglected.

(b) Lessen the efficiency of the **packaging** through differential thermal expansion or cracking or melting of the radiation shielding material; or

(c) In combination with moisture, accelerate corrosion.

544. Except as required in para. 515 for a **package** transported by air, a **package** shall be so designed that, under the ambient condition specified in para. 545, the temperature of the accessible surfaces of a **package** shall not exceed 50°C, unless the **package** is transported under **exclusive use**.

545. In applying paras 543 and 544, the ambient temperature shall be assumed to be 38°C.

546. In applying para. 543, the solar insolation conditions shall be assumed to be as specified in Table XII.

547. A **package** which includes thermal protection for the purpose of satisfying the requirements of the thermal test specified in para. 628 shall be so designed that such protection will remain effective if the **package** is subjected to the tests specified in paras 619–624 and 627(a) and (b) or 627(b) and (c), as appropriate. Any such protection on the exterior of the **package** shall not be rendered ineffective by conditions likely to be encountered in routine handling or transport, or in accidents, and which are not simulated in the tests referred to above, e.g. by ripping, cutting, skidding, abrasion or rough handling.

548. A **package** shall be so designed that, if it were subjected to:

(a) The tests specified in paras 619–624, it would restrict the loss of **radioactive contents** to not more than 10^{-6} **A$_2$** per hour; and

(b) The tests specified in paras 626, 627(b), 628 and 629 and the test in paras

    (i) 627(c), when the **package** has a mass not greater than 500 kg, an overall density not greater than 1000 kg/m^3 based on the external dimensions, and **radioactive contents** greater than 1000 A_2 not as **special form radioactive material**, or

    (ii) 627(a), for all other **packages**,

it would restrict the accumulated loss of **radioactive contents** in a period of one week to not more than 10 A_2 for krypton-85 and not more than A_2 for all other radionuclides.

Where mixtures of different radionuclides are present, the provisions of paras 304–306 shall apply except that for krypton-85 an effective A_2 value equal to 100 TBq (2000 Ci) may be used. For case (a) above, the evaluation shall take into account the external contamination limitations of paras 408 and 409.

Requirements for Type B(U) packages

549. **Type B(U) packages** shall meet the requirements for **Type B packages** specified in paras 541–548, and the requirements specified in paras 550–556.

550. A **package** for irradiated nuclear fuel with activity greater than 37 PBq (10^6 Ci) shall be so designed that, if it were subjected to the water immersion test specified in para. 630, there would be no rupture of the **containment system**.

551. Compliance with the permitted activity release limits shall depend neither upon filters nor upon a mechanical cooling system.

552. A **package** shall not include a pressure relief system from the **containment system** which would allow the release of **radioactive material** to the environment under the conditions of the tests specified in paras 619–624 and 626–629.

553. A **package** shall be so designed that if it were at the **maximum normal operating pressure** and it were subjected to the tests specified in paras 619–624 and 626–629, the level of strains in the **containment system** would not attain values which would adversely affect the **package** in such a way that it would fail to meet the applicable requirements.

554. A **package** shall not have a **maximum normal operating pressure** in excess of a gauge pressure of 700 kPa (7 kgf/cm^2).

555. Except as required in para. 515 for a **package** transported by air, the maximum temperature of any surface readily accessible during transport of a **package** shall not exceed 85°C in the absence of insolation under the ambient condition specified in para. 545; and the **package** shall be carried under **exclusive use**, as specified in para. 544, if this maximum temperature exceeds 50°C. Account may be taken of barriers or screens to give protection to transport workers without the need for the barriers or screens being subject to any test.

556. A **package** shall be designed for an ambient temperature range from −40°C to +38°C.

Requirements for Type B(M) packages

557. **Type B(M) packages** shall meet the requirements for **Type B packages** specified in paras 541–548, except that for **packages** to be transported solely within a specified country or solely between specified countries, conditions other than those given in paras 545, 546 and 556 above may be assumed with the approval of the **competent authorities** of these countries. As far as practicable, the requirements for **Type B(U) packages** specified in paras 550–556 shall be met.

558. Intermittent venting of **Type B(M) packages** may be permitted during transport, provided that the operational controls for venting are acceptable to the relevant **competent authorities**.

REQUIREMENTS FOR PACKAGES CONTAINING FISSILE MATERIAL

559. Except as provided in para. 560, **packages** containing **fissile material** shall be so designed, and used, to comply with the requirements specified in paras 561–568, as well as those specified in paras 518–520, 524 or 541, as applicable, taking into account the nature, activity and form of the contents.

560. **Packages** meeting one of the requirements of paras 560(a)–560(f) shall be excepted from the requirements specified in paras 561–568, and from the other requirements of these Regulations that apply specifically to **fissile material**; such **packages**, however, shall be regulated as non-**fissile radioactive material packages** as applicable, and shall still be subject to those requirements of these Regulations which pertain to their radioactive nature and properties.

(a) **Packages** containing individually not more than 15 g of **fissile material**, provided that the smallest external dimension of each **package** is not less than 10 cm. For unpackaged material, the quantity limitation shall apply to the **consignment** being carried in or on the **conveyance**.

(b) **Packages** containing homogeneous hydrogenous solutions or mixtures satisfying the conditions listed in Table XIII. For unpackaged material, the quantity limitations in Table XIII shall apply to the **consignment** being carried in or on the **conveyance**.

(c) **Packages** containing uranium enriched in uranium-235 to a maximum of 1% by mass, and with a total plutonium and uranium-233 content not exceeding 1% of the mass of uranium-235, provided that the **fissile material** is distributed essentially homogeneously throughout the material. In addition, if uranium-235 is present in metallic, oxide or carbide forms, it shall not form a lattice arrangement within the **package**.

(d) **Packages** containing not more than 5 g of **fissile material** in any 10 litre volume, provided that the **radioactive material** is contained in **packages**

TABLE XIII. LIMITATIONS ON HOMOGENEOUS HYDROGENOUS SOLUTIONS OR MIXTURES OF FISSILE MATERIAL

Parameters	Uranium-235 only	Any other **fissile material** (including mixtures)
Minimum H/X[a]	5200	5200
Maximum concentration of **fissile material** (g/L)	5	5
Maximum mass of **fissile material** in a **package** or **conveyance** (g)	800[b]	500

[a] Where H/X is the ratio of the number of hydrogen atoms to the number of atoms of fissile nuclide.

[b] With a total plutonium and uranium-233 content of not more than 1% of the mass of uranium-235.

which will maintain the limitations on **fissile material** distribution under conditions likely to be encountered during routine transport.

(e) **Packages** containing individually not more than 1 kg of total plutonium, of which not more than 20% by mass may consist of plutonium-239, plutonium-241, or any combination of those radionuclides.

(f) **Packages** containing liquid solutions of uranyl nitrate enriched in uranium-235 to a maximum of 2% by mass, with a total plutonium and uranium-233 content not exceeding 0.1% of the mass of uranium-235, and with a minimum nitrogen to uranium atomic ratio (N/U) of 2.

561. **Packages** containing **fissile material** shall be transported and stored in accordance with the relevant controls in Section IV.

562. **Fissile material** shall be packaged and shipped in such a manner that subcriticality is maintained under conditions likely to be encountered during normal conditions of transport and in accidents. The following contingencies shall be considered:

(a) Water leaking into or out of **packages**;
(b) The loss of efficiency of built-in neutron absorbers or moderators;
(c) Possible rearrangement of the **radioactive contents** either within the **package** or as a result of loss from the **package**;
(d) Reduction of spaces between **packages** or **radioactive contents**;
(e) **Packages** becoming immersed in water or buried in snow; and
(f) Possible effects of temperature changes.

563. A **packaging** for **fissile material** shall be so designed that, if it were subjected to the tests specified in paras 619–624:

(a) Neither the volume nor any spacing on the basis of which nuclear criticality control for the purpose of para. 567(a) has been assessed would suffer more than 5% reduction, and the construction of the **packaging** would prevent the entry of a 10 cm cube; and

(b) Water would not leak into or out of any part of the **package** unless water in-leakage into or out-leakage from that part, to the optimum foreseeable extent, has been assumed for the purposes of paras 566 and 567; and

(c) The configuration of the **radioactive contents** and the geometry of the **containment system** would not be altered so as to increase the neutron multiplication significantly.

Undamaged and damaged packages

564. For the purposes of the evaluation in this subsection:

(a) Undamaged shall mean the condition of the **package** as it is designed to be presented for transport;

(b) Damaged shall mean the evaluated or demonstrated condition of the **package** if it had been subjected to whichever of the following combination of tests is the more limiting:

   (i) The tests specified in paras 619–624 followed by the tests specified in paras 626–628 and completed by the tests specified in paras 631–633. The mechanical test of para. 627 shall be that required by para. 548.

   (ii) The tests specified in paras 619–624 followed by the test in para. 629.

Individual packages in isolation

565. In determining the subcriticality of individual **packages** in isolation, it shall be assumed that water can leak into or out of all void spaces of the **package**, including those within the **containment system**. However, if the **design** incorporates special features to prevent such leakage of water into or out of certain void spaces, even as a result of human error, absence of leakage may be assumed in respect of those void spaces. Special features shall include the following:

(a) Multiple high standard water barriers, each of which would remain leaktight if the **package** were damaged (see para. 564); a high degree of quality control in the production and maintenance of **packagings**; and special tests to demonstrate the closure of each **package** before **shipment**; or

(b) Other features given **multilateral approval**.

566. The individual **package** damaged or undamaged shall be subcritical under the conditions specified in paras 564 and 565, taking into account the physical and chemical characteristics including any change in those characteristics which could occur when the **package** is damaged and with the conditions of moderation and reflection as specified below:

(a) For the material within the **containment system**: the material arranged in the **containment system**
   (i) In the configuration and moderation that results in maximum neutron multiplication; and
   (ii) With close reflection of the **containment system** by water 20 cm thick (or equivalent) or such greater reflection of the **containment system** as may additionally be provided by the surrounding material of the **packaging**; and, in addition
(b) If any part of the material escapes from the **containment system**: that material arranged in
   (i) The configuration and moderation that results in maximum neutron multiplication; and
   (ii) With close reflection of that material by water 20 cm thick (or equivalent).

Arrays of packages

567. An array of **packages** shall be subcritical. A number 'N' shall be derived assuming that if **packages** were stacked together in any arrangement with the stack closely reflected on all sides by water 20 cm thick (or its equivalent) both of the following conditions would be satisfied:

(a) Five times 'N' undamaged **packages** without anything between the **packages** would be subcritical; and
(b) Two times 'N' damaged **packages** with hydrogenous moderation between **packages** to the extent which results in the greatest neutron multiplication would be subcritical.

Subcriticality evaluation assumptions

568. In evaluating the subcriticality of **fissile material** in its transport configuration, the following shall apply:

(a) The determination of subcriticality for irradiated **fissile material** may be based on the actual irradiation experience, taking into account significant variations in composition.

(b) For irradiated **fissile material** of unknown irradiation experience the following assumptions shall be made in determining subcriticality:
   (i) If its neutron multiplication decreases with irradiation, the material shall be regarded as unirradiated;
   (ii) If its neutron multiplication increases with irradiation, the material shall be regarded as irradiated to the point corresponding to the maximum neutron multiplication; and
(c) For unspecified **fissile material**, such as residues or scrap, whose fissile composition, mass, concentration, moderation ratio or density is not known or cannot be identified, the assumption shall be made in determining subcriticality that each parameter that is not known has the value which gives the maximum neutron multiplication under credible conditions of transport.

Section VI
TEST PROCEDURES

DEMONSTRATION OF COMPLIANCE

601. Demonstration of compliance with the performance standards required in Section V shall be accomplished by any of the methods listed below or by a combination thereof.

(a) Performance of tests with specimens representing **LSA-III, special form radioactive material** (solid **radioactive material** or capsules), or with prototypes or samples of the **packaging**, where the contents of the specimen or the **packaging** for the tests shall simulate as closely as practicable the expected range of **radioactive contents**, and the specimen or **packaging** to be tested shall be prepared as normally presented for transport.

(b) Reference to previous satisfactory demonstrations of a sufficiently similar nature.

(c) Performance of tests with models of appropriate scale incorporating those features which are significant with respect to the item under investigation, when engineering experience has shown results of such tests to be suitable for design purposes. When a scale model is used, the need for adjusting certain test parameters, such as penetrator diameter or compressive load, shall be taken into account.

(d) Calculation, or reasoned argument, when the calculation procedures and parameters are generally agreed to be reliable or conservative.

602. After the specimen, prototype or sample has been subjected to the tests, appropriate methods of assessment shall be used to assure that the requirements of this section have been fulfilled in conformance with the performance and acceptance standards prescribed in Section V.

TEST FOR LSA-III MATERIAL

603. Solid material representing no less than the entire contents of the **package** shall be immersed for 7 days in water at ambient temperature. The volume of water to be used in the test shall be sufficient to ensure that at the end of the 7 day test period, the free volume of the unabsorbed and unreacted water remaining shall be at least 10% of the volume of the solid test sample itself. The water shall have an initial pH of 6–8 and a maximum conductivity of 1 mS/m (10 μmho/cm) at 20°C. The total activity of the free volume of water shall be measured following the 7 day immersion of the test sample.

TESTS FOR SPECIAL FORM RADIOACTIVE MATERIAL

General

604. The tests which shall be performed on specimens that comprise or simulate **special form radioactive material** are: the impact test the percussion test, the bending test and the heat test.

605. A different specimen may be used for each of the tests.

606. After each test specified in paras 607–611, a leaching assessment or volumetric leakage test shall be performed on the specimen by a method no less sensitive than the methods given in para. 612 for indispersable solid material and para. 613 for encapsulated material.

Test methods

607. *Impact test:* The specimen shall drop onto the target from a height of 9 m. The target shall be as defined in para. 618.

608. *Percussion test:* The specimen shall be placed on a sheet of lead which is supported by a smooth solid surface and struck by the flat face of a steel billet so as to produce an impact equivalent to that resulting from a free drop of 1.4 kg through 1 m. The flat face of the billet shall be 25 mm in diameter with the edges rounded off to a radius of (3.0 ± 0.3) mm. The lead, of hardness number 3.5 to 4.5 on the Vickers scale and not more than 25 mm thick, shall cover an area greater than that covered by the specimen. A fresh surface of lead shall be used for each impact. The billet shall strike the specimen so as to cause maximum damage.

609. *Bending test:* The test shall apply only to long, slender sources with both a minimum length of 10 cm and a length to minimum width ratio of not less than 10. The specimen shall be rigidly clamped in a horizontal position so that one half of its length protrudes from the face of the clamp. The orientation of the specimen shall be such that the specimen will suffer maximum damage when its free end is struck by the flat face of a steel billet. The billet shall strike the specimen so as to produce an impact equivalent to that resulting from a free vertical drop of 1.4 kg through 1 m. The flat face of the billet shall be 25 mm in diameter with the edges rounded off to a radius of (3.0 ± 0.3) mm.

610. *Heat test:* The specimen shall be heated in air to a temperature of 800°C and held at that temperature for a period of 10 minutes and shall then be allowed to cool.

611. Specimens that comprise or simulate **radioactive material** enclosed in a sealed capsule may be excepted from:

(a) The tests prescribed in paras 607 and 609 provided they are alternatively subjected to the Class 4 impact test prescribed in the International Organization for Standardization document ISO 2919-1980(E), "Sealed radioactive sources — Classification"; and
(b) The test prescribed in para. 610 provided they are alternatively subjected to the Class 6 temperature test specified in the International Organization for Standardization document ISO 2919-1980(E), "Sealed radioactive sources — Classification".

Leaching and volumetric leakage assessment methods

612. For specimens which comprise or simulate indispersable solid material, a leaching assessment shall be performed as follows:

(a) The specimen shall be immersed for 7 days in water at ambient temperature. The volume of water to be used in the test shall be sufficient to ensure that, at the end of the 7 day test period, the free volume of the unabsorbed and unreacted water remaining shall be at least 10% of the volume of the solid test sample itself. The water shall have an initial pH of 6–8 and a maximum conductivity of 1 mS/m (10 μmho/cm) at 20°C.
(b) The water with specimen shall then be heated to a temperature of (50 ± 5)°C and maintained at this temperature for 4 hours.
(c) The activity of the water shall then be determined.
(d) The specimen shall then be stored for at least 7 days in still air of relative humidity not less than 90% at 30°C.
(e) The specimen shall then be immersed in water of the same specification as in (a) above and the water with the specimen heated to (50 ± 5)°C and maintained at this temperature for 4 hours.
(f) The activity of the water shall then be determined.

613. For specimens which comprise or simulate **radioactive material** enclosed in a sealed capsule, either a leaching assessment or a volumetric leakage assessment shall be performed as follows:

(a) The leaching assessment shall consist of the following steps:
  (i) The specimen shall be immersed in water at ambient temperature. The water shall have an initial pH of 6–8 with a maximum conductivity of 1 mS/m (10 μmho/cm) at 20°C.
  (ii) The water and specimen shall be heated to a temperature of (50 ± 5)°C and maintained at this temperature for 4 hours.
  (iii) The activity of the water shall then be determined.

(iv) The specimen shall then be stored for at least 7 days in still air at a temperature not less than 30°C.

(v) The process in (i), (ii) and (iii) shall be repeated.

(b) The alternative volumetric leakage assessment shall comprise any of the tests prescribed in the International Organization for Standardization document ISO/TR 4826-1979(E), "Sealed radioactive sources — Leak test methods", which are acceptable to the **competent authority**.

TESTS FOR PACKAGES

Preparation of a specimen for testing

614. All specimens shall be examined before testing in order to identify and record faults or damage including the following:

(a) divergence from the **design**;
(b) defects in construction;
(c) corrosion or other deterioration; and
(d) distortion of features.

615. The **containment system** of the **package** shall be clearly specified.

616. The external features of the specimen shall be clearly identified so that reference may be made simply and clearly to any part of such specimen.

Testing the integrity of the containment system and shielding and evaluating criticality safety

617. After the applicable tests specified in paras 619–633:

(a) Faults and damage shall be identified and recorded;
(b) It shall be determined whether the integrity of the **containment system** and shielding has been retained to the extent required in Section V for the **packaging** under test; and
(c) For **packages** containing **fissile material**, it shall be determined whether the assumptions made in paras 562–567 regarding the most reactive configuration and degree of moderation of the fissile contents, of any escaped material, and of one or more **packages** are valid.

Target for drop tests

618. The target for the drop tests specified in paras 607, 622, 625(a) and 627 shall be a flat, horizontal surface of such a character that any increase in its resistance to

TABLE XIV. FREE DROP DISTANCE FOR TESTING PACKAGES TO NORMAL CONDITIONS OF TRANSPORT

Package mass (kg)	Free drop distance (m)
package mass < 5 000	1.2
5 000 ≤ package mass < 10 000	0.9
10 000 ≤ package mass < 15 000	0.6
15 000 ≤ package mass	0.3

displacement or deformation upon impact by the specimen would not significantly increase the damage to the specimen.

Tests for demonstrating ability to withstand normal conditions of transport

619. The tests are: the water spray test, the free drop test, the stacking test and the penetration test. Specimens of the **package** shall be subjected to the free drop test, the stacking test and the penetration test, preceded in each case by the water spray test. One specimen may be used for all the tests, provided that the requirements of para. 620 are fulfilled.

620. The time interval between the conclusion of the water spray test and the succeeding test shall be such that the water has soaked in to the maximum extent, without appreciable drying of the exterior of the specimen. In the absence of any evidence to the contrary, this interval shall be taken to be two hours if the water spray is applied from four directions simultaneously. No time interval shall elapse, however, if the water spray is applied from each of the four directions consecutively.

621. *Water spray test:* The specimen shall be subjected to a water spray test that simulates exposure to rainfall of approximately 5 cm per hour for at least one hour.

622. *Free drop test:* The specimen shall drop onto the target so as to suffer maximum damage in respect of the safety features to be tested.

(a) The height of drop measured from the lowest point of the specimen to the upper surface of the target shall be not less than the distance specified in Table XIV for the applicable mass. The target shall be as defined in para. 618.

(b) For **packages** containing **fissile material** the free drop test specified above shall be preceded by a free drop from a height of 0.3 m on each corner or, in the case of a cylindrical package, onto each of the quarters of each rim.

(c) For rectangular fibreboard or wood **packages** not exceeding a mass of 50 kg, a separate specimen shall be subjected to a free drop onto each corner from a height of 0.3 m.

(d) For cylindrical fibreboard **packages** not exceeding a mass of 100 kg, a separate specimen shall be subjected to a free drop onto each of the quarters of each rim from a height of 0.3 m.

623. *Stacking test:* Unless the shape of the **packaging** effectively prevents stacking, the specimen shall be subjected, for a period of 24 h, to a compressive load equal to the greater of the following:

(a) The equivalent of 5 times the mass of the actual **package**; and

(b) The equivalent of 13 kPa (0.13 kgf/cm^2) multiplied by the vertically projected area of the **package**.

The load shall be applied uniformly to two opposite sides of the specimen, one of which shall be the base on which the **package** would normally rest.

624. *Penetration test:* The specimen shall be placed on a rigid, flat, horizontal surface which will not move significantly while the test is being carried out.

(a) A bar of 3.2 cm in diameter with a hemispherical end and a mass of 6 kg shall be dropped and directed to fall, with its longitudinal axis vertical, onto the centre of the weakest part of the specimen, so that, if it penetrates sufficiently far, it will hit the **containment system**. The bar shall not be significantly deformed by the test performance.

(b) The height of drop of the bar measured from its lower end to the intended point of impact on the upper surface of the specimen shall be 1 m.

Additional tests for Type A packages designed for liquids and gases

625. A specimen or separate specimens shall be subjected to each of the following tests unless it can be demonstrated that one test is more severe for the specimen in question than the other, in which case one specimen shall be subjected to the more severe test.

(a) *Free drop test:* The specimen shall drop onto the target so as to suffer the maximum damage in respect of containment. The height of the drop measured from the lowest part of the specimen to the upper surface of the target shall be 9 m. The target shall be as defined in para. 618.

(b) *Penetration test:* The specimen shall be subjected to the test specified in para. 624 except that the height of drop shall be increased to 1.7 m from the 1 m specified in para. 624(b).

Tests for demonstrating ability to withstand accident conditions in transport

626. The specimen shall be subjected to the cumulative effects of the tests specified in para. 627 and para. 628, in that order. Following these tests, either this specimen or a separate specimen shall be subjected to the effect(s) of the water immersion test(s) as specified in para. 629 and, if applicable, para. 630.

627. *Mechanical test:* The mechanical test consists of three different drop tests. Each specimen shall be subjected to the applicable drops as specified in para. 548. The order in which the specimen is subjected to the drops shall be such that, on completion of the mechanical test, the specimen shall have suffered such damage as will lead to the maximum damage in the thermal test which follows.

(a) For drop I, the specimen shall be dropped onto the target so as to suffer the maximum damage, and the height of the drop measured from the lowest point of the specimen to the upper surface of the target shall be 9 m. The target shall be as defined in para. 618.

(b) For drop II, the specimen shall be dropped so as to suffer the maximum damage onto a bar rigidly mounted perpendicularly on the target. The height of the drop measured from the intended point of impact of the specimen to the upper surface of the bar shall be 1 m. The bar shall be of solid mild steel of circular section, (15.0 ± 0.5) cm in diameter, and 20 cm long unless a longer bar would cause greater damage, in which case a bar of sufficient length to cause maximum damage shall be used. The upper end of the bar shall be flat and horizontal with its edges rounded off to a radius of not more than 6 mm. The target on which the bar is mounted shall be as described in para. 618.

(c) For drop III, the specimen shall be subjected to a dynamic crush test by positioning the specimen on the target so as to suffer maximum damage by the drop of a 500 kg mass from 9 m onto the specimen. The mass shall consist of a solid mild steel plate 1 m by 1 m and shall fall in a horizontal attitude. The height of the drop shall be measured from the underside of the plate to the highest point of the specimen. The target on which the specimen rests shall be as defined in para. 618.

628. *Thermal test:* The thermal test shall consist of the exposure of a specimen fully engulfed, except for a simple support system, in a hydrocarbon fuel/air fire of sufficient extent and in sufficiently quiescent ambient conditions to provide an average emissivity coefficient of at least 0.9, with an average flame temperature of at least 800°C for a period of 30 minutes, or shall be any other thermal test which provides the equivalent total heat input to the **package**. The fuel source shall extend horizontally at least 1 m, and shall not extend more than 3 m, beyond any external surface of the specimen, and the specimen shall be positioned 1 m above the surface of the fuel source. After the cessation of external heat input, the specimen shall not be

cooled artificially and any combustion of materials of the specimen shall be allowed to proceed naturally. For demonstration purposes, the surface absorptivity coefficient shall be either 0.8 or that value which the **package** may be demonstrated to possess if exposed to the fire specified; and the convective coefficient shall be that value which the designer can justify if the **package** were exposed to the fire specified. With respect to the initial conditions for the thermal test, the demonstration of compliance shall be based upon the assumption that the **package** is in equilibrium at an ambient temperature of 38°C. The effects of solar radiation may be neglected prior to and during the tests, but must be taken into account in the subsequent evaluation of the **package** response.

629. *Water immersion test:* The specimen shall be immersed under a head of water of at least 15 m for a period of not less than eight hours in the attitude which will lead to maximum damage. For demonstration purposes, an external gauge pressure of at least 150 kPa (1.5 kgf/cm^2) shall be considered to meet these conditions.

Water immersion test for packages containing irradiated nuclear fuel

630. The specimen shall be immersed under a head of water of at least 200 m for a period of not less than one hour. For demonstration purposes, an external gauge pressure of at least 2 MPa (20 kgf/cm^2) shall be considered to meet these conditions.

Water leakage test for packages containing fissile material

631. **Packages** for which water in-leakage or out-leakage to the extent which results in greatest reactivity has been assumed for purposes of assessment under paras 564–567 shall be excepted from the test.

632. Before the specimen is subjected to the water leakage test specified below, it shall be subjected to the tests in para. 627(b), and either para. 627(a) or (c) as required by para. 548, and the test specified in para. 628.

633. The specimen shall be immersed under a head of water of at least 0.9 m for a period of not less than eight hours and in the attitude for which maximum leakage is expected.

Section VII

APPROVAL AND ADMINISTRATIVE REQUIREMENTS

GENERAL

701. **Competent authority** approval shall be required for the following:

(a) **Special form radioactive material** (see paras 702 and 703);
(b) All **packages** containing **fissile material** (see paras 710–712, 713 and 714);
(c) **Type B packages** — **Type B(U)** and **Type B(M)** (see paras 704–709, 713 and 714);
(d) **Special arrangements** (see paras 720–722);
(e) Certain **shipments** (see paras 716–719);
(f) Radiation protection programme for special use **vessels** (see para. 472); and
(g) Calculation of unlisted A_1 and A_2 values (see para. 302).

APPROVAL OF SPECIAL FORM RADIOACTIVE MATERIAL

702. The **design** for **special form radioactive material** shall require **unilateral approval**. An application for approval shall include:

(a) A detailed description of the **radioactive material** or, if a capsule, the contents; particular reference shall be made to both physical and chemical states;
(b) A detailed statement of the **design** of any capsule to be used;
(c) A statement of the tests which have been done and their results, or evidence based on calculative methods to show that the **radioactive material** is capable of meeting the performance standards, or other evidence that the **special form radioactive material** meets the applicable requirements of the Regulations; and
(d) Evidence of a quality assurance programme.

703. The **competent authority** shall establish an approval certificate stating that the approved **design** meets the requirements for **special form radioactive material** and shall attribute to that **design** an identification mark. The certificate shall specify the details of the **special form radioactive material**.

APPROVAL OF PACKAGE DESIGNS

Approval of Type B(U) package designs

704. Each **Type B(U) package design** shall require **unilateral approval**, except that a **package design** for **fissile material**, which is also subject to paras 710–712, shall require **multilateral approval**.

705. An application for approval shall include:

(a) A detailed description of the proposed **radioactive contents** with particular reference to their physical and chemical states and the nature of the radiation emitted;

(b) A detailed statement of the **design**, including complete engineering drawings and schedules of materials and methods of construction to be used;

(c) A statement of the tests which have been done and their results, or evidence based on calculative methods or other evidence that the **design** is adequate to meet the applicable requirements;

(d) The proposed operating and maintenance instructions for the use of the **packaging**;

(e) If the **package** is designed to have a **maximum normal operating pressure** in excess of 100 kPa (1.0 kgf/cm^2) gauge, the application for approval shall, in particular, state, in respect of the materials of construction of the **containment system**, the specifications, the samples to be taken, and the tests to be made;

(f) Where the proposed **radioactive contents** are irradiated fuel, the applicant shall state and justify any assumption in the safety analysis relating to the characteristics of the fuel;

(g) Any special stowage provisions necessary to ensure the safe dissipation of heat from the **package**; consideration shall be given to the various modes of transport to be used and type of **conveyance** or **freight container**;

(h) A reproducible illustration not larger than 21 cm × 30 cm showing the make-up of the **package**; and

(i) Evidence of a quality assurance programme.

706. The **competent authority** shall establish an approval certificate stating that the **design** meets the requirements for **Type B(U) packages**.

Approval of Type B(M) package designs

707. Each **Type B(M) package design**, including those for **fissile material** which are also subject to paras 710–712, shall require **multilateral approval**.

708. An application for approval of a **Type B(M) package design** shall include, in addition to the information required in para. 705 for **Type B(U) packages**:

(a) A list of the specific requirements for **Type B(U) packages** specified in para. 549, with which the **package** does not conform;
(b) Any proposed supplementary operational controls to be applied during transport not routinely provided for in these Regulations, but which are necessary to ensure the safety of the **package** or to compensate for the deficiencies listed in (a) above, such as human intervention for temperature or pressure measurements or for periodic venting, taking into account the possibility of unexpected delay.
(c) A statement relative to any restrictions on the mode of transport and to any special loading, carriage, unloading or handling procedures; and
(d) The maximum and minimum ambient conditions (temperature, solar radiation) expected to be encountered during transport and which have been taken into account in the **design**.

709. The **competent authority** shall establish an approval certificate stating that the **design** meets the applicable requirements for **Type B(M) packages**.

Approval of package designs for fissile material

710. Each **package design** for **fissile material**, which is not excepted according to para. 560 from the requirements that apply specifically to **packages** containing **fissile material**, shall require **multilateral approval**.

711. An application for approval shall include all information necessary to satisfy the **competent authority** that the **design** meets the requirements of para. 559 and evidence of a quality assurance programme.

712. The **competent authority** shall establish an approval certificate stating that the **design** meets the requirements of paras 561–568.

Approvals under the 1967, 1973 and the 1973 (As Amended) Editions of the Regulations

713. **Packagings** manufactured to a **design** approved by the **competent authority** under the provisions of the 1967 Edition of these Regulations may continue to be used, subject to **multilateral approval**. Changes in the **design** of the **packaging** or in the nature or quantity of the authorized **radioactive contents** which, as determined by the **competent authority**, would significantly affect safety shall be required to meet the 1985 Edition of the Regulations. No new construction of such **packagings**

shall be permitted to commence. A serial number according to the provision of para. 438 shall be assigned to and marked on the outside of each **packaging**.

714. **Packagings** manufactured to a **design** approved under the provisions of the 1973 Edition and the 1973 (As Amended) Edition of these Regulations may continue to be used until 31 December 1992.

After this date:

(a) **Multilateral approval** shall be required;
(b) A serial number, according to the provisions of para. 438, shall be assigned to and marked on the outside of each **packaging**.

Changes in the **design** of the **packaging** or in the nature or quantity of the authorized **radioactive contents** which, as determined by the **competent authority**, would significantly affect safety shall be required to meet the 1985 Edition of the Regulations. Each Member State shall require that all **packagings** for which construction begins after 31 December 1995 meet the 1985 Edition of the Regulations in full.

NOTIFICATION AND REGISTRATION OF SERIAL NUMBERS

715. The **competent authority** shall be informed of the serial number of each **packaging** manufactured to a **design** approved under paras 704, 707, 710, 713 and 714. The **competent authority** shall maintain a register of such serial numbers.

APPROVAL OF SHIPMENTS

716. Except as allowed in para. 717, **multilateral approval shall be required for:**

(a) The **shipment** of **Type B(M) packages** especially designed to allow controlled intermittent venting;
(b) The **shipment** of **Type B(M) packages** containing **radioactive material** with an activity greater than 3×10^3 A_1 or 3×10^3 A_2, as appropriate, or 1000 TBq (20 kCi), whichever is the lower;
(c) The **shipment** of **packages** containing **fissile materials** if the sum of the **transport indexes** of the individual **packages** exceeds 50 as provided in para. 465; and
(d) Radiation protection programmes for **shipments** by special use **vessels** according to para. 472.

717. A **competent authority** may authorize transport into or through its country without **shipment** approval, by a specific provision in its **design** approval (see para. 723).

718. An application for **shipment** approval shall include:

(a) The period of time, related to the **shipment**, for which the approval is sought;
(b) The actual **radioactive contents**, the expected modes of transport, the type of **conveyance**, and the probable or proposed route; and
(c) The details of how the special precautions and special administrative or operational controls, referred to in the **package design** approval certificates issued under paras 706, 709 and 712 are to be put into effect.

719. Upon approval of the **shipment**, the **competent authority** shall issue an approval certificate.

APPROVAL OF SHIPMENT UNDER SPECIAL ARRANGEMENT

720. Each **consignment** shipped under **special arrangement** shall require **multilateral approval**.

721. An application for approval of a **shipment** under **special arrangement** shall include all the information necessary to satisfy the **competent authority** that the overall level of safety in transport is at least equivalent to that which would be provided if all the applicable requirements of the Regulations had been met. The application shall also include:

(a) A statement of the respects in which, and of the reasons why, the **consignment** cannot be made in full accordance with the applicable requirements; and
(b) A statement of any special precautions or special administrative or operational controls which are to be employed during transport to compensate for the failure to meet the applicable requirements.

722. Upon approval of a **shipment** under **special arrangement**, the **competent authority** shall issue an approval certificate.

COMPETENT AUTHORITY APPROVAL CERTIFICATES

723. Four types of approval certificates may be issued: **special form radioactive material**, **special arrangement**, **shipment** and **package design**. The **package design** and **shipment** approval certificates may be combined into a single certificate.

Competent authority identification marks

724. Each approval certificate issued by a **competent authority** shall be assigned an identification mark. The mark shall be of the following generalized type:

VRI/Number/Type Code

(a) VRI represents the international vehicle registration identification code of the country issuing the certificate.

(b) The number shall be assigned by the **competent authority**, and shall be unique and specific with regard to the particular **design** or **shipment**. The **shipment** approval identification mark shall be clearly related to the **design** approval identification mark.

(c) The following type codes shall be used in the order listed to indicate the types of approval certificates issued:

    AF   **Type A package design** for **fissile material**
    B(U)   **Type B(U) package design** [B(U)F if for **fissile material**]
    B(M)   **Type B(M) package design** [B(M)F if for **fissile material**]
    IF   **Industrial package design** for **fissile material**
    S   **Special form radioactive material**
    T   **Shipment**
    X   **Special arrangement**.

(d) For **package design** approval certificates, other than those issued under the provisions of paras 713 or 714, the symbols '-85' shall be added to the type code of the **package design**.

725. These type codes shall be applied as follows:

(a) Each certificate and each **package** shall bear the appropriate identification mark, comprising the symbols prescribed in para. 724(a), (b), (c) and (d) above, except that, for **packages**, only the applicable **design** type codes including, if applicable, the symbols '-85', shall appear following the second stroke, that is, the 'T' or 'X' shall not appear in the identification marking on the **package**. Where the **design** approval and **shipment** approval are combined, the applicable type codes do not need to be repeated. For example:

    A/132/B(M)F-85:   A **Type B(M) package design** approved for **fissile material**, requiring **multilateral approval**, for which the **competent authority** of Austria has assigned the **design** number 132 (to be marked on both the **package** and on the **package design** approval certificate);

A/132/B(M)F-85T: The **shipment** approval issued for a **package** bearing the identification mark elaborated above (to be marked on the certificate only)

A/137/X-85: A **special arrangement** approval issued by the **competent authority** of Austria, to which the number 137 has been assigned (to be marked on the certificate only);

A/139/IF-85: An **industrial package design** for **fissile material** approved by the **competent authority** of Austria, to which **package design** number 139 has been assigned (to be marked on both the **package** and on the **package design** approval certificate).

(b) Where **multilateral approval** is effected by validation, only the identification mark issued by the country of origin of the **design** or **shipment** shall be used. Where **multilateral approval** is effected by issue of certificates by successive countries, each certificate shall bear the appropriate mark and the **package** whose **design** was so approved shall bear all appropriate identification marks. For example:

   A/132/B(M)F-85
   CH/28/B(M)F-85

would be the identification mark of a **package** which was originally approved by Austria and was subsequently approved, by separate certificate, by Switzerland. Additional identification marks would be tabulated in a similar manner on the **package**.

(c) The revision of a certificate shall be indicated by a parenthetical expression following the identification mark on the certificate. For example, A/132/B(M)F-85(Rev.2) would indicate revision 2 of the Austrian **package design** approval certificate; or A/132/B(M)F-85(Rev.0) would indicate the original issuance of the Austrian **package design** approval certificate. For original issuances, the parenthetical entry is optional and other words such as 'original issuance' may also be used in place of 'Rev.0'. Certificate revision numbers may only be issued by the country issuing the original approval certificate.

(d) Additional symbols (as may be necessitated by national requirements) may be added in brackets to the end of the identification mark; for example, A/132/B(M)F-85(SP503).

(e) It is not necessary to alter the identification mark on the **packaging** each time that a revision to the **design** certificate is made. Such re-marking shall be made only in those cases where the revision to the **package design** certificate involves a change in the letter type codes for the **package design** following the second stroke.

CONTENTS OF APPROVAL CERTIFICATES

Special form radioactive material approval certificates

726. Each approval certificate issued by a **competent authority** for **special form radioactive material** shall include the following information:

(a) Type of certificate.
(b) The **competent authority** identification mark.
(c) The issue date and an expiry date.
(d) List of applicable national and international regulations, including the edition of the IAEA Regulations for the Safe Transport of Radioactive Material under which the **special form radioactive material** is approved.
(e) The identification of the **special form radioactive material**.
(f) A description of the **special form radioactive material**.
(g) **Design** specifications for the **special form radioactive material** which may include references to drawings.
(h) A specification of the **radioactive contents** which includes the activities involved and which may include the physical and chemical form.
(i) A specification of the applicable quality assurance programme as required in para. 209.
(j) If deemed appropriate by the **competent authority**, reference to the identity of the applicant.
(k) Signature and identification of the certifying official.

Special arrangement approval certificates

727. Each approval certificate issued by a **competent authority** for a **special arrangement** shall include the following information:

(a) Type of certificate.
(b) The **competent authority** identification mark.
(c) The issue date and an expiry date.
(d) Mode(s) of transport.
(e) Any restrictions on the modes of transport, type of **conveyance**, **freight container** and any necessary routing instructions.
(f) List of applicable national and international regulations, including the edition of the IAEA Regulations for the Safe Transport of Radioactive Material under which the **special arrangement** is approved.
(g) The following statement:

> "This certificate does not relieve the consignor from compliance with any requirement of the government of any country through or into which the package will be transported."

(h) References to certificates for alternative **radioactive contents**, other **competent authority** validation, or additional technical data or information, as deemed appropriate by the **competent authority**.

(i) Description of the **packaging** by a reference to the drawings or a specification of the **design**. If deemed appropriate by the **competent authority**, a reproducible illustration not larger than 21 cm by 30 cm showing the make-up of the **package** should also be provided, accompanied by a very brief description of the **packaging** including materials of construction, gross mass, general outside dimensions and appearance.

(j) A brief specification of the authorized **radioactive contents**, including any restrictions on the **radioactive contents** which might not be obvious from the nature of the **packaging**. This shall include the physical and chemical forms, the activities involved (including those of the various isotopes, if appropriate), amounts in grams (for **fissile material**), and whether **special form radioactive material**.

(k) Additionally, for **package designs** for **fissile material**:

    (i) a detailed description of the authorized **radioactive contents**;
    (ii) the value of the **transport index** for nuclear criticality control;
    (iii) any special features, on the basis of which the absence of water from certain void spaces has been assumed in the criticality assessment; and
    (iv) any determination (based on para. 568(a)) on which decreased neutron multiplication is assumed in the criticality assessment as a result of actual irradiation experience.

(l) A detailed listing of any supplementary operational controls required for preparation, loading, transport, stowage, unloading and handling of the **consignment**, including any special stowage provisions for the safe dissipation of heat.

(m) If deemed appropriate by the **competent authority**, reasons for the **special arrangement**.

(n) Description of the compensatory measures to be applied as a result of the **shipment** being under **special arrangement**.

(o) Reference to information provided by the applicant relating to the use of the **packaging** or specific actions to be taken prior to the **shipment**.

(p) A statement regarding the ambient conditions assumed for purposes of **design** if these are not in accordance with those specified in paras 545, 546 and 556, as applicable.

(q) Any emergency arrangements deemed necessary by the **competent authority**.

(r) A specification of the applicable quality assurance programme as required in para. 209.

(s) If deemed appropriate by the **competent authority**, reference to the identity of the applicant and to the identity of the **carrier**.

(t) Signature and identification of the certifying official.

Shipment approval certificates

728. Each approval certificate for a **shipment** issued by a **competent authority** shall include the following information:

(a) Type of certificate.
(b) The **competent authority** identification mark.
(c) The issue date and an expiry date.
(d) List of applicable national and international regulations, including the edition of the IAEA Regulations for the Safe Transport of Radioactive Material under which the **shipment** is approved.
(e) Any restrictions on the modes of transport, type of **conveyance**, **freight container**, and any necessary routing instructions.
(f) The following statement:

> "This certificate does not relieve the consignor from compliance with any requirement of the government of any country through or into which the package will be transported."

(g) A detailed listing of any supplementary operational controls required for preparation, loading, transport, stowage, unloading, and handling of the **consignment**, including any special stowage provisions for the safe dissipation of heat.
(h) Reference to information provided by the applicant relating to specific actions to be taken prior to **shipment**.
(i) Reference to the applicable **design** approval certificate.
(j) A brief specification of the actual **radioactive contents**, including any restrictions on the **radioactive contents** which might not be obvious from the nature of the **packaging**. This shall include the physical and chemical forms, the total activities involved (including those of the various isotopes, if appropriate), amounts in grams (for **fissile material**), and whether **special form radioactive material**.
(k) Any emergency arrangements deemed necessary by the **competent authority**.
(l) A specification of the applicable quality assurance programme as required in para. 209.
(m) If deemed appropriate by the **competent authority**, reference to the identity of the applicant.
(n) Signature and identification of the certifying official.

Package design approval certificates

729. Each approval certificate of the **design** of a **package** issued by a **competent authority** shall include the following information:

(a) Type of certificate.
(b) The **competent authority** identification mark.
(c) The issue date and an expiry date.
(d) Any restriction on the modes of transport, if appropriate.
(e) List of applicable national and international regulations, including the edition of the IAEA Regulations for the Safe Transport of Radioactive Material under which the **design** is approved.
(f) The following statement:

"This certificate does not relieve the consignor from compliance with any requirement of the government of any country through or into which the package will be transported."

(g) References to certificates for alternative **radioactive contents**, other **competent authority** validation, or additional technical data or information, as deemed appropriate by the **competent authority**.
(h) A statement authorizing **shipment** where **shipment** approval is required under para. 716, if deemed appropriate.
(i) Identification of the **packaging**.
(j) Description of the **packaging** by a reference to the drawings or **specification** of the **design**. If deemed appropriate by the **competent authority**, a reproducible illustration not larger than 21 cm by 30 cm showing the make-up of the **package** should also be provided, accompanied by a very brief description of the **packaging** including materials of construction, gross mass, general outside dimensions and appearance.
(k) Specification of the **design** by reference to the drawings.
(l) A brief specification of the authorized **radioactive content**, including any restrictions on the **radioactive contents** which might not be obvious from the nature of the **packaging**. This shall include the physical and chemical forms, the activities involved (including those of the various isotopes, if appropriate), amounts in grams (for **fissile material**), and whether **special form radioactive material**.
(m) Additionally, for **packages** for **fissile material**:
  (i) A detailed description of the authorized **radioactive contents**;
  (ii) The value of the **transport index** for nuclear criticality control;
  (iii) Any special features, on the basis of which the absence of water from certain void spaces has been assumed in the criticality assessment; and
  (iv) Any determination (based on para. 568(a)) on which decreased neutron multiplication is assumed in the criticality assessment as a result of actual irradiation experience.
(n) For **Type B(M) packages**, a statement specifying those prescriptions of paras. 550–556 with which the **package** does not conform and any amplifying information which may be useful to other **competent authorities**.

(o) A detailed listing of any supplementary operational controls required for preparation, loading, transport, stowage, unloading and handling of the **consignment**, including any special stowage provisions for the safe dissipation of heat.

(p) Reference to information provided by the applicant relating to the use of the **packaging** or specific actions to be taken prior to **shipment**.

(q) A statement regarding the ambient conditions assumed for purposes of **design** if these are not in accordance with those specified in paras 545, 546 and 556, as applicable.

(r) A specification of the **quality assurance** programme as required in para. 209.

(s) Any emergency arrangements deemed necessary by the **competent authority**.

(t) If deemed appropriate by the **competent authority**, reference to the identity of the applicant.

(u) Signature and identification of the certifying official.

VALIDATION OF CERTIFICATES

730. **Multilateral approval** may be by validation of the original certificate issued by the **competent authority** of the country of origin of the **design** or **shipment**. Such validation may take the form of an endorsement on the original certificate or the issuance of a separate endorsement, annex, supplement, etc., by the **competent authority** of the country through or into which the **shipment** is made.

Appendix I

EXCERPTS FROM LIST OF UNITED NATIONS NUMBERS, PROPER SHIPPING NAME AND DESCRIPTION AND SUBSIDIARY RISKS

Number	Name and description	Subsidiary risks
2910	RADIOACTIVE MATERIAL, EXCEPTED PACKAGE, — INSTRUMENTS or ARTICLES, — LIMITED QUANTITY OF MATERIAL, — ARTICLES MANUFACTURED FROM NATURAL URANIUM or DEPLETED URANIUM or NATURAL THORIUM, — EMPTY PACKAGING	
2912	RADIOACTIVE MATERIAL, LOW SPECIFIC ACTIVITY (LSA), N.O.S.[a]	
2913	RADIOACTIVE MATERIAL, SURFACE CONTAMINATED OBJECTS (SCO)	
2918	RADIOACTIVE MATERIAL, FISSILE, N.O.S.[a]	
2974	RADIOACTIVE MATERIAL, SPECIAL FORM, N.O.S.[a]	
2975	THORIUM METAL, PYROPHORIC	Liable to spontaneous combustion
2976	THORIUM NITRATE, SOLID	Oxidizing substance
2977	URANIUM HEXAFLUORIDE, FISSILE containing more than 1.0 per cent uranium-235	Corrosive
2978	URANIUM HEXAFLUORIDE, fissile excepted or non-fissile	Corrosive
2979	URANIUM METAL, PYROPHORIC	Liable to spontaneous combustion
2980	URANYL NITRATE HEXAHYDRATE SOLUTION	Corrosive
2981	URANYL NITRATE, SOLID	Oxidizing substance
2982	RADIOACTIVE MATERIAL, N.O.S.[a]	

[a] N.O.S. — not otherwise specified.

Appendix II
CONVERSION FACTORS AND PREFIXES

This edition of the Regulations for the Safe Transport of Radioactive Material uses, as primary units, the International System of Units (SI). In some cases however, subsidiary units which have been traditionally used are shown in parentheses following the primary units to assist users. As a result, the values which are controlling are those with SI units; the values with subsidiary units are only approximations thereof. The conversion factors for the dually specified units are:

RADIATION UNITS

Activity in becquerel (Bq) or curie (Ci)

$1 \text{ Ci} = 3.7 \times 10^{10} \text{ Bq}$

Dose equivalent in sievert (Sv) or rem

$1 \text{ rem} = 1.0 \times 10^{-2} \text{ Sv}$

PRESSURE

Pressure in pascal (Pa) or (kgf/cm^2)

$1 \text{ kgf/cm}^2 = 9.806 \times 10^4 \text{ Pa}$

CONDUCTIVITY

Conductivity in siemens per metre (S/m) or (mho/cm)

$10 \ \mu\text{mho/cm} = 1 \text{ mS/m}$

or

$1 \text{ mho/cm} = 100 \text{ S/m}$

SI PREFIXES

The prefixes to be used with the SI units are:

Multiplying factor	Prefix	Symbol
$1\ 000\ 000\ 000\ 000\ 000\ 000 = 10^{18}$	exa	E
$1\ 000\ 000\ 000\ 000\ 000 = 10^{15}$	peta	P
$1\ 000\ 000\ 000\ 000 = 10^{12}$	tera	T
$1\ 000\ 000\ 000 = 10^{9}$	giga	G
$1\ 000\ 000 = 10^{6}$	mega	M
$1\ 000 = 10^{3}$	kilo	k
$100 = 10^{2}$	hecto	h
$10 = 10^{1}$	deka	da
$0.1 = 10^{-1}$	deci	d
$0.01 = 10^{-2}$	centi	c
$0.001 = 10^{-3}$	milli	m
$0.000\ 001 = 10^{-6}$	micro	μ
$0.000\ 000\ 001 = 10^{-9}$	nano	n
$0.000\ 000\ 000\ 001 = 10^{-12}$	pico	p
$0.000\ 000\ 000\ 000\ 001 = 10^{-15}$	femto	f
$0.000\ 000\ 000\ 000\ 000\ 001 = 10^{-18}$	atto	a

CONTRIBUTORS TO DRAFTING AND REVIEW

Abe, H.	Central Research Institute of Electric Power Industry, Japan
Altemos, E.	International Civil Aviation Organization
Araki, S.	Science and Technology Agency, Japan
Baekelandt, L.	Nationale Instelling voor Radioactif Afval en Splijtstoffen/ Organisme national des déchets radioactifs et des matières fissiles, Belgium
Barker, R.	International Atomic Energy Agency
Barrett, L.	US Department of Energy, United States of America
Biaggio, A.L.	Comisión Nacional de Energía Atómica, Argentina
Blackman, D.	Department of Transport, United Kingdom
Blum, P.	Transnucléaire, France
Brélaz, P.	Bundesamt für Energiewirtschaft, Switzerland
Burbidge, G.	Nordion International Inc., Canada
Cheshire, R.D.	British Nuclear Fuels plc, United Kingdom
Collin, F.W.	Physikalisch-Technische Bundesanstalt, Federal Republic of Germany
Cosack, M.	Physikalisch-Technische Bundesanstalt, Federal Republic of Germany
Devillers, C.	Commissariat à l'énergie atomique, France
Dicke, G.	Ontario Hydro, Canada
Dufva, B.	Swedish Nuclear Power Inspectorate, Sweden
Dybeck, P.	Swedish Nuclear Fuel and Waste Management Co., Sweden
Ek, P.	Swedish Nuclear Power Inspectorate, Sweden

Ershov, V.N.	All-Union Project and Research Institute of Complex Power Technology, Union of Soviet Socialist Republics
Eyre, P.	Atomic Energy Control Board, Canada
Faloci, C.	Comitato Nazionale per l'Energia Nucleare, Italy
Fasten, C.	Staatliches Amt für Atomsicherheit und Strahlenschutz, German Democratic Republic
Fedin, V.I.	Ministry of Public Health, Union of Soviet Socialist Republics
Gibson, W.H.	Atomic Energy of Canada Ltd, Canada
Gioria, G.	Comitato Nazionale per la Ricerca e per lo Sviluppo dell'Energia Nucleare e delle Energie Alternative, Italy
Golder, F.	Institute of Isotopes of the Hungarian Academy of Sciences, Hungary
Goldfinch, E.P.	Central Electricity Generating Board, United Kingdom
Grenier, M.	Commissariat à l'énergie atomique, France
Hamard, J.	Commissariat à l'énergie atomique, France
Hamos, A.	International Atomic Energy Agency
Harmon, L.	US Department of Energy, United States of America
Hladík, I.	Czechoslovak Atomic Energy Commission, Czechoslovakia
Hopkins, D.R.	Nuclear Regulatory Commission, United States of America
Jackson, B.J.	Nordion International Inc., Canada
Jankowski, G.J.	Amersham International Ltd, United Kingdom
Johnson, G.M.	International Air Transport Association
Joseph, D.	Atomic Energy Control Board, Canada

Kafka, G.	Bundesministerium für Öffentliche Wirtschaft und Verkehr, Austria
Kitamura, T.	Power Reactor and Nuclear Fuel Development Corporation, Japan
Koponen, H.	International Atomic Energy Agency
Kubo, M.	Power Reactor and Nuclear Fuel Development Corporation, Japan
Levin, I.	International Atomic Energy Agency
Luna, R.E.	Sandia National Laboratories, United States of America
Marchal, M.A.	Commission of the European Communities
Matsui, H.	Japan Atomic Energy Research Institute, Japan
Mayr, K.	Österreichisches Forschungszentrum Seibersdorf GmbH, Austria
McLellan, J.J.	Atomic Energy Control Board, Canada
Nagahama, H.	Kobe Steel Ltd, Japan
Nakajima, T.	Science and Technology Agency, Japan
Neubauer, J.	Österreichisches Forschungszentrum Seibersdorf GmbH, Austria
Nitsche, F.	Institut für Energetik, German Democratic Republic
Nomura, M.	Japan Atomic Energy Research Institute, Japan
Olsson, R.	Swedish Nuclear Power Inspectorate, Sweden
Orsini, A.	Comitato Nazionale per la Ricerca e per lo Sviluppo dell'Energia Nucleare e delle Energie Alternative, Italy
O'Sullivan, R.A.	Department of Transport, United Kingdom
Patek, P.	Österreichisches Forschungszentrum Seibersdorf GmbH, Austria

Pettersson, B.G.	Swedish Nuclear Power Inspectorate, Sweden
Pittuck, A.	Transport Canada, Canada
Pope, R.B.	International Atomic Energy Agency
Ridder, K.	Bundesministerium für Verkehr, Federal Republic of Germany
Ringot, C.	Commissariat à l'énergie atomique, France
Rojas De Diego, J.	International Atomic Energy Agency
Roosemont, G.	Ministère de la santé publique et de l'environnement, Belgium
Sannen, M.	Transnubel, Belgium
Sanui, T.	Nuclear Fuel Transport Co. Ltd, Japan
Scott, H.	Nuclear Regulatory Commission, United States of America
Selling, H.A.	Ministry of Housing, Physical Planning and the Environment, Netherlands
Shaw, K.B.	National Radiological Protection Board, United Kingdom
Shiomi, S.	Central Research Institute of Electric Power Industry, Japan
Singh, D.	Atomic Energy Regulatory Board, India
Siwicki, R.	Central Laboratory for Radiation Protection, Poland
Smith, L.	Bundesamt für Energiewirtschaft, Switzerland
Smith, R.	British Nuclear Fuels plc, United Kingdom
Stalder, F.	Eidgenössisches Institut für Reaktorforschung, Switzerland
Svahn, B.	National Institute of Radiation Protection, Sweden
Takeshita, K.	Transnuclear Ltd, Japan

Van Oosterwijk, R.	Ministerie van Verkeer en Waterstaat, Netherlands
Venet, M.	International Federation of Air Line Pilots Associations
Wangler, M.E.	US Department of Transportation, United States of America
Wardelmann, H.	International Maritime Organization
Warden, D.	Amersham International Ltd, United Kingdom
Wieser, K.	Bundesanstalt für Materialforschung und -prüfung, Federal Republic of Germany
Yasogawa, Y.	Japan Marine Surveyors and Sworn Measurers Association, Japan
Young, C.N.	Department of Transport, United Kingdom

Advisory Group Meeting

Vienna, Austria: 13–17 January 1986

Review Panel Meetings

Vienna, Austria: 22–26 June 1987,
10–14 July 1989

SELECTION OF IAEA PUBLICATIONS RELATING TO THE SAFE TRANSPORT OF RADIOACTIVE MATERIAL

SAFETY SERIES

6	Regulations for the Safe Transport of Radioactive Material, 1985 Edition (As Amended 1990)	1990
7	Explanatory Material for the IAEA Regulations for the Safe Transport of Radioactive Material (1985 Edition), Second Edition (As Amended 1990)	1990
9	Basic Safety Standards for Radiation Protection, 1982 Edition	1982
37	Advisory Material for the IAEA Regulations for the Safe Transport of Radioactive Material (1985 Edition), Third Edition (As Amended 1990)	1990
80	Schedules of Requirements for the Transport of Specified Types of Radioactive Material Consignments (As Amended 1990)	1990
87	Emergency Response Planning and Preparedness for Transport Accidents Involving Radioactive Material	1988

TECDOC SERIES

287	INTERTRAN: A System for Assessing the Impact from Transporting Radioactive Material	1983
295	Directory of Transport Packaging Test Facilities	1983
318	Transport of Radioactive Materials by Post	1984
374	Discussion of and Guidance on the Optimization of Radiation Protection in the Transport of Radioactive Material	1986
375	International Studies on Certain Aspects of the Safe Transport of Radioactive Materials, 1980–1985	1986
398	Assessment of the Radiological Impact of the Transport of Radioactive Materials	1986

399	Assessment of the Application of the IAEA Regulations for the Safe Transport of Radioactive Material	1986
413	Competent Authority Regulatory Control of the Transport of Radioactive Material	1987
423	Recommendations for Providing Protection during the Transport of Uranium Hexafluoride	1987
552	Directory of National Competent Authorities' Approval Certificates for Package Design and Shipment of Radioactive Material, 1990 Edition *(replaces IAEA-TECDOC-422)*	1990

TRAINING COURSE SERIES

1	Safe Transport of Radioactive Material	1990

INDEX

Accident conditions: 103, 134, 527, 626–629

Activity limits: Section III, 301, 311, 314, Table IV, Table VI
  A_1: 110, 134(c) and (d), 301–306, Table I, Table II, 312, Table IV, 456(a), 701(g), 716(b)
  A_2: 110, 131, 134(c) and (d), 301–306, Table I, Table II, 312, Table IV, Table VI, 456(a), 501, 539, 540, 548, 701(g), 716(b)

Air (transport by): 102, 125(c), 134, 433, 473–475, 477(b), 515–518, 524, 541, 544, 555

Ambient conditions: 515, 516, 543–545, 556, 603, 612, 613, 628, 708(d), 727(p), 729(q)

Approval certificate: 313, 315, 402(b), 447(n), 463, 703, 706, 709, 712, 718(c), 719, 722–729

Approval — general: Section VII

Approval — multilateral: 113, 141, 211, 302, 466(f), 565(b), 704, 707, 710, 713, 714(a), 716, 720, 725(a) and (b), 730

Approval — special arrangement: 720–723, 724(c), 725(a), 727

Approval — unilateral: 114, 402(c), 702, 704

Basic Safety Standards: 201–204

Carrier: 107, 112, 115, 210, 453, 454, 727(r)

Categories: 435–445, Table IX, Table X, 447, Fig. 2, Fig. 3, Fig. 4, 461, 470(a), 479

Competent authority: 107, 113, 114, 116, 117, 141, 203, 209–211, 302, 410, 438(a), 447(n), 454–458, 463, 464, 472(a), 482, 484, 503(c), 529, 557, 558, 613(b), 701, 703, 706, 709, 711–715, 717, 719, 721–730

Compliance assurance: 107, 117, 210

Consignee: 118, 128, Table XI(a), 483, 484

Consignment: 107, 112, 113, 118–120, 133, 140, 141, 206, 208, 211, 405, 406, Table III, 430, 432, 444, Fig. 5, 447, 448, 451, 453(c), 454–457, 462, 465(a), Table XI(b), 467–469, 472, 473, 476, 477, 481, 560(a) and (b), 720, 721(a), 727(l), 728(g), 729(o)

Consignor: 107, 119, 120, 128, 133, 209, 210, 405, Table VIII, 431(c), 446–459, Table XI(a), 477, 484, 727(g), 728(f), 729(f)

Containment: 134, 516, 543, 625(c)

107

Containment system: 121, 132, 134, 147, 208, 401(a) and (b), 402(d), 516, 530–534, 536, 539(c), 543, 550, 552, 553, 563(c), 565, 566, 615, 617, 624(a), 625(a), 705(e)

Contamination: 122–124, 144, 408–414, Table III, 417, 421(c), 425, 548
  Fixed: 123, 144, 413
  Non-fixed: 123, 124, 144, 408–409, Table III, 413, 417, 421(c), 425(c)

Conveyance: 125, 128, 130, 146, 204(b), 311, Table III, 410, 412–414, 425, 427, Table VI, 453(b), 465, Table XI, 480, 505, 560(a) and (b), Table XIII, 705(g), 718(b), 727(e), 728(e)
  Activity limit: 311, 427, Table VI
  TI limits: 311, 427, Table VI, Table XI

Cooling system: 474, 551

Criticality: 146, 429, Table VIII, 431, 563(a), 727(k), 729(m)

Customs: 108, 483

Dangerous goods: 105, 406, 407, 460(b), 478(b), 519, 521(b)

Deck area: 125(b), 126, Table XI

Decontamination: 413

Dose limitation *(see also* **Radiation control***):* 203

Emergency: 207, 453(c), 727(q), 728(k), 729(s)

Empty packaging: 421, 452, 477(c)

Excepted: 134(a), 308–310, Table III, 414–420, Table IV, 421, 443, 447(l) and (m), 452, 472, 481, 540, 560, 611, 631

Exclusive use: 128, 146, 405, 414, 425(b), Table V, Table VIII, 432–434, 435(c) and (d), Table IX, 444, 447, 466–469, 471, 473, 475, 544

Exemption: 402(c)

Fabrication: 103, 529

Filter: 551

Fissile *(See also* **Criticality***):* 129, 131(a)(iii), Section III, 314, 315, 401(b) and (c), 402(b), 415(c), 423, Table VIII, 431, 442(b), 447(i) and (m), 457(e), 464, Table XI, 482, 559–568, Table XIII, 617(c), 622(b), 631–633, 701(b), 704, 707, 710–712, 716(c), 724(c), 725(a), 727(j) and (k), 728(j), 729(l) and (m)

Freight container: 128, 130, 133–135, 146, 409, Table III, 424, 428, Table VIII, Table X, 440–444, 447(o), 453(a), 460, 465, Table XI, 467, 470(a), 478–480, 521–523, 705(g), 727(e), 728(e)

Gas: 131(b), 138, 145, 147, Table IV, Table V, Table VI, 522, 533, 540, 625

Heat: 401(b), 453(a), 463, 503(b), 543, 604, 610, 628, 705(g), 727(l), 728(g), 729(o)

Identification mark: 438(a), 447(n), 457(a), 703, 724, 725, 726(b), 727(b), 728(b), 729(b)

Industrial package: (*See* **Package — industrial**)

Insolation (solar radiation): 515, 546, Table XII, 555

Inspection: 107, 136, 209, 210, 401, 402

Label: 421(d), 440–443, 445, Fig. 2, Fig. 3, Fig. 4, 452, 467, 470(a)

Leaching: 131(c), 503(c), 606, 612, 613

Leakage (loss): 410, 503(c), 517, 535, 539(b), 565, 606, 612, 613, 631–633

Low specific activity (LSA): 131, 134(b), 146, Table I, 311, 403, 414, 422–427, Table V, Table VI, 428, Table VIII, 442(a), 444, 447(e), 465(a), 468, 481, 501, 522, 601(a), 603

Maintenance: 103, 209, 210, 565(a), 705(d)

Marking: 407, 418(b), 419(b), 436–439, 446, 725(a)

Mass: 143, 150, 315(a), 436, 442(b), 447(i), 457(e), 505, 507, 548(b), 560, Table XIII, 568(c), 622, Table XIV, 623(a), 624(a), 627(c), 727(i), 729(j)

Maximum normal operating pressure: 132, 553, 554, 705(e)

Normal conditions: 103, 134, 411, 521(c), 527, 543, 555, 562, 619–624, Table XIV

Notification: 113, 455–458, 715

Number 'N': 429, 567

Operational controls: 132, 474, 558, 708(b), 718(c), 721(b), 727(l), 728(g), 729(o)

Other dangerous properties: 105, 407, 440, 514

Overpack: 130, 133, 146, 409, Table III, 428, Table VIII, 431–435, Table X, 440, 441, 442(c), 447(o), 453(a), 460, 461, 463, 465, 466, Table XI, 467, 469, 470(a), 478–480

Package design: 209, 313, 315, 437, 438(c), 439, 447(n), 455, 704–714, 718(c), 723, 724(c) and (d), 725(a),(c) and (e), 727(k), 729

Package — excepted: 134(a), 308–310, Table III, 415–420, Table IV, 421, 443, 447(l), 452

Package — industrial: 134(b), 311, Table V, 518–523, 724(c), 725(a)

Package — Type A: 110, 134(c), 312, 437, 524–540, 625, 724(c)

Package — Type B: 134(d), 313, 401(b), 402, 539, 541–549, 557, 701(c)

Package — Type B(M): 438(c), 439, 456(b), 473, 474, 557–558, 701(c), 707–709, 716(a) and (b), 724(c), 725(a), 729(n)

Package — Type B(U): 438(c), 439, 456(a), 549–557, 701(c), 704–706, 708, 724(c)

Packaging: 103, 107, 121, 127, 130, 131(c), 134, 135, 138, 209, 210, 314, 315, 401(b) and (c), 402(b), 403, 415(b), 421, 426, 436–438, 438(b), 452, 477(c) and (e), Section V, 505–514, 519, 528, 532, 536, 539, 543(b), 563, 565(a), 566(a), 601(a), 617(b), 623, 705, 713–715, 725(e), 727(i),(j) and (o), 728(j), 729(i),(j),(l) and (p)

Placard: 407, 443–446, Fig. 5, Fig. 6, 467, 468

Placarding: *(See* **Placard***)*

Post: 310, 415(d), 476, 477

Pressure: 132, 147, 401(a), 402(c), 517, 521(b), 530, 534, 535, 552–554, 629, 630, 705(e), 708(b)

Pressure relief: 535, 552

Quality assurance: 107, 136, 209, 702(d), 705(i), 711, 726(i), 727(r), 728(l), 729(r)

Radiation control: 428, Table VIII, 460(a), 478(a), 483

Radiation exposure: 137, 146, 201–204, 206, 428, Table VIII, 460(a), 478(a), 483

Radiation level: 137, 204, 205, 311, 410, 413, 416, 418(a), 422, 428(a), 431(c), 432–434, 435(a) and (d), Table IX, 465(b), 469, 470(b), 471, 475, 519(b), 521(c), 523(b), 537(b), 542

Radiation protection: 201–205, 472(a), 701(f), 716(d)

Radiation shield: 135, 536, 543(b)

Radioisotopic cardiac pacemaker: 104(b)

Rail (transport by): 125(a), 145, 151, 433, 467–470, 475

Responsibility: 108, 202, 210, 446

Road (transport by): 125, 145, 151, 433, 467–470, 475

Segregation: 205, 206, 406, 460, 461

Serial number: 438(b), 713, 714(b), 715

Shielding: 131, 134, 135, 401(b), 425(a), 426, 519(b), 521(c), 523(b), 537(b), 542, 543, 617

Shipment: 113, 140, 141, 209, Section IV, 401, 402, 447(n), 447(p), 455, 456, 457(b), 458, 459, 464, 469(a)(iii), 472(a), 482, 565(a), 701(e), 716–724, 725(a) and (b), 727(n) and (o), 728, 729(h) and (p), 730

Shipping: 447(a),(c) and (l), 448

Special arrangement: 141, 146, 211, 433, 435(e) and (f), 447(n), 456(c), 464, 471, 475, 482, 701(d), 720–723, 724(c), 725(a), 727

Special form: 110, 127, 134(c) and (d), 142, 304(a), 312(a), Table IV, 447(h) and (n), 457(d), 502–504, 531, 548(b), 601(a), 604–613, 701(a), 702–703, 723, 724(c), 726, 727(j), 728(j), 729(l)

Specific activity *(See also* **Low specific activity***)*: 131, 139, 143

Storage in transit: 103, 146, 209, 211, Section IV, Table XI(a), 478–482

Stowage: 126, 133, 210, 453, 463, 472, 705, 727(l), 728(g), 729(o)

Surface contaminated objects (SCO): 134(b), 144, 146, Table I, 311, 403, 414, 422–427, Table V, Table VI, 428, Table VIII, 442(a)(ii), 444, 447(f), 468

Tank: 134(b),(c) and (d), 135, 145, 146, 404, 424, 428, Table VIII, 440, 441, 442(c), 443, 444, 453(a), 460, 465, Table XI(b) and (d), 467, 470, 478–480, 521–523

Tank container: 145, 440, 521(c), 522, 523(b)

Tank vehicle: 145

Temperature: 132, 402(c), 515, 516, 528, 538, 544–556, 562(f), 603, 610, 611(b), 612, 613(a), 628, 708(b) and (d)

Tests (mechanical and thermal): 134, 401(c), 402(d), 501, 503, 519, 521(b), 523(b), 537, 539(a), 540, 542, 543, 547, 548(a) and (b), 550, 552, 553, 555, 563, 564(b), 565(a), Section VI, 702(c), 705(c) and (e)

Tie-down: 145, 527

Transport documents: 120, 442(c), 447, 448, 453

Transport index (TI): 146, 428–431, Table VIII, 432, 435(a), (b) and (c), Table IX, Table X, 442(d), 447(k), 464, 465(a), 466, Table XI, 469, 479, 480, 482, 716(c), 727(k), 729(m)

Transport workers: 201–205, 555

Ullage: 538

Uncompressed gas: 147, 540

United Nations Numbers: 444, Fig. 5, Fig. 6, 447(b),(d) and (l), 468, Appendix I

Unpackaged: 130, 146, 418(a), 425, 426, Table V(a), 428, Table VIII, 444, 468, 560(a) and (b)

Vehicle: 125(a), 126, 145, 151, Table XI, 467–471, 724(a)

Venting: 132, 558, 708(b), 716(a)

Vessel: 125(b), 126, 152, 433, Table XI, 471, 472, 701(f), 716(d)

Water: 102, 105, 125(b), 131, 208, Table VI, 439, 501, 503(c), 509, 550, 562(a) and (e), 563(b), 565, 566(a) and (b), 567, 603, 612(a),(b), (c) and (e), 613(a), 619–621, 626, 629–633, 727(k), 729(m)

White label: *(See* **Categories***)*

Yellow label: *(See* **Categories***)*